What Utility Safety Leaders Do

Tips, Tactics, Strategies, and Insights for Leaders

Matthew A. Forck, CSP and JLW

SAFESTRAT, LLC

Copyright © 2015 Matthew A. Forck
All rights reserved.
No part of this book may be reproduced, stored in a retrieval system or transmitted by any means, electronic, mechanical, photocopying, recording or otherwise, without the written permission from the author.

ISBN Number 978-1-5056-7048-6

Printed in the United States of America

Published by
SafeStrat, LLC
Safety strategies... for LIFE!

Cover and book interior design by Susan Ferber, FerberDesign

This book is dedicated to ALL utility workers and those family members who support them—we don't do heart surgery, but they can't do it without us!

Contents

Preface.. viii

Making the Bed Should Be a Safety Rule
What Leaders Know about Wheel Chocks and Keystone Habits........................... *1*

The Safest Damn Utility in the Country.. *5*

Safety Culture: What's under Your Helmet?.. *9*

Scoring 100% Safe Work
A Leadership Secret to Success in the Utility Industry *13*

How to Slow Down and Avoid Injuries on the Job.. *15*

Leading with Chaos . . . and It Is All Chaos!
Four Steps to Help You Lead in Times of Confusion, Pressure, and Disorder *18*

Thinking Differently about Target Zero *23*

Unreasonableness: How Leaders Make Progress *26*

Macho Safety
The New Definition of Toughness *30*

What Toothpaste and Safety Leadership Have in Common *34*

How to Create Your Safety Vision.. *37*

Leadership
Thinking about What Others Think... *40*

How Leaders Think about Near Misses
Turning Remote Misses into Near Misses *44*

How Safety Leaders Win.. *49*

Pashtunwali: A Safety Code
Are We Really Our Brother's Keeper?.. *52*

How Leaders Move from Soft Skills to SMART Skills....................................... *56*

You Can't Pick the Time.. *61*

Always Take the Shark Training
Five Things Linemen Should always Do *63*

Four Key Steps to Conducting an Effective Job Briefing *67*

Finding a MAP for Safety Committees Success!... 70

The S.T.O.R.M. Model
One Approach to Effective Near Miss Reporting... 73

The Space Between
What's in Your Space?..76

Small Stuff Matters
So Everyone Finishes Safe! 80

Safety's Broken Windows
Why Wheel Chocks and Steel-Toed Shoes Really Matter! 83

Safety Is NOT the Right Thing to Do!... 86

Five Keys to Leading Lineman Safety
Getting Respect and Getting Results... 89

Is Your Culture Killing You?
How to Move from a Culture of Honor to a Culture of Safety 92

Is Training All Wrong?
What Leaders Recognize about Training... 95

How are Your Executive Safety Skills?... 100

Hazard Intelligence
Four Intelligences That Can Change Your Safety Culture 104

Finding the Gorillas!
How Leaders Deal with Attention Blindness... 108

Do You Have Safety With-It-ness?...113

Chicken Rings
What's in Your Circle of Safety? 116

Check Down
How Football Quarterback Can Make Us Better at Safety 119

Unreasonable Leadership
Challenging What We Think is True about Safety Performance...................................... 123

Safety's Innovation Cycle
What Safety Leaders Understand about Innovation.. 128

The Secrets to $1,000 an Hour Work
What Leaders Know about Spending Time and the Value It Brings................................. 134

What Day Will You Get Hurt?
It's about an Attitude, Not a Day *139*

How Can Safety Leaders Sleep at Night?
They Know These Seven Keys to Worker Engagement! *143*

Yield, for Safety's Sake *149*

Seven Strategies for Improving the Safety Record of Your Crew *151*

Constant Safety Awareness
Five Keys to Managing Your Space Between *154*

Five Hidden Safety Secrets of Line Work *157*

The S.T.O.R.M. Model for Near Miss Reporting *160*

Sounds of Silence
What's Wrong with Near Miss Reporting *162*

Will You Be Ready When "IT" Happens? *167*

Random Acts of Safety Kindness *169*

Oh Yeah, and Then There Is Safety Leadership! *172*

In Closing *175*

About the Author *176*

Preface

""Your best thinking got you here . . . "

Welcome to *What Utility Safety Leaders Do—Tips, Tactics, Strategies, and Insights for Leaders* . . .

Between my junior and senior year of college at Truman State University in Kirksville, Missouri, I was lucky to land a temporary laborer job with the local utility company. My job was to paint and cut grass. I worked alone. It was the longest summer of my life. What I didn't know at the time, was that this summer, the summer of 1991, was a prelude to my career. Not the 'boring' part . . . the utility piece!

After graduating with academic honors a year later, I prepared myself for a high profile job at the state or federal level. After an exhaustive job search I had exactly one job offer . . . a ditch digger with the same utility that I had worked for the summer prior. I was broke from college and needed the cash—and digging ditches paid well. I took it. After six months with a shovel in my hand I actually moved up, meter reader. From there I took an overheard distribution electrical line apprenticeship in 1995 and earned a journey card as a distribution lineman in 1997.

Growing up a 'sports guy' I loved being a lineman. There are not many occupations where you can have the same trust, respect, and dynamics that are fostered on a highly competitive sports team. As a line worker, I spent time working difficult jobs with the 'guys' day and night. We struggled together in mud, rain, ice, snow, and heat. We climbed poles, gloved 12 KV, sticked 69 KV, and tapped transformers. We moved heavy equipment, operate tractors, backhoes, trenchers, and more. We shared a beer or two (off the clock). Barbecued with family, helped each other move, and watched each other's children grow. It was rewarding. I still treasure those experiences and relationships. I learned about safety.

The utility business is one of those rare industries where just one little mistake, one at-risk act, can end one's hopes, dreams, goals, and ambitions—forever lost. Utility injuries are unforgiving.

Preface

Because of the risks in the utility industry, and the harsh reality of the life changing nature of utility incidents, we need safety leadership; and we need it like never before. Utilities are pretty good at rules. We have rules for chocking truck tires, PPE, confined space entry, trenching and shoring, job planning, fall prevention, and the list goes on and on. And, we need each and every one of these rules. But, the utility industry is also very autonomous. Since there are never two jobs exactly the same, the work rules must always be interpreted, applied, conditions change and rules must be reapplied. To that end, safety leadership is the key to long-term success... and safety leadership is NOT outlined in our safety manuals.

Over a decade ago, I left the line crews and began working as a safety superior; I served over 400 line workers, substation technicians, and natural gas workers in outstate Missouri. I went back to the books and earned my board certification in safety, Certified Safety Professional, or CSP. I was able to learn from and listen to utility workers both in my normal assigned area and those across the country. I saw first hand what happens when there is a lack of leadership... I spent time in emergency rooms... I spoke with families after their loved ones suffered significant injuries.

This book is written for each utility worker, his/her supervisor, manager, director, and safety professional because what got us here (to this point) won't get us there (to the next level of safety excellence). To make it to zero injuries and zero at-risk acts we need to think differently. This book has nearly 50 chapters. Each chapter is an independent article on safety leadership. The goal is to offer different tips, tactics, insights, and strategies. You will see some common themes and stories, but pay attention to those as examples of leadership at work, themes to consider, and key elements to utility safety. I hope the chapters make you think, and that you find a number of new ideas from each section.

Thank you for working in the utility industry. What this industry does is the foundation of today's economy. The utility industry, and those men and women who dedicate their lives to it, provide the standard of living we enjoy. Work safe. Lead.

1 Making the Bed Should Be a Safety Rule

What Leaders Know about Wheel Chocks and Keystone Habits

Did you know that many researches that study happiness and effectiveness claim that making your bed every morning is the one single habit that is correlated with better productivity?

The cell phone rang; being in a meeting, I ignored it. It immediately rang again. I picked up. It was the regional dispatcher. I can still remember his words, "Electrical contact Matt, we've got two men down."

Once on site, I found that the crew had been setting poles and laying out phases to reconductor a three mile section of line. The six man crew, with over 100 years of experience between them, was going to work one last pole then go home for the weekend. Given the experience of the crew and the fact that this job was normally done with three men, not six, it was a cake job for a Friday.

The only obvious and major hazards on the job was a 12,470 volt phase-to-phase overhead line and the busy two lane highway. The crew knowingly positioned their truck under the line to avoid setting up on a busy road. Shortly after starting work, the boom contacted the overhead line as the men were pulling material off the truck. Both men touching the truck when the boom contacted the line received an electrical contact. One man died.

After the fact, when I was able to look at the pictures and begin writing the incident report, I started by just listing the safety rules that could have been employed to prevent this—truck grounding, boom spotter, covering the phases with line hose, improved job planning and hazard recognition, staying off the truck—and the list goes on. But, there is one rule that the crew did not follow that is glaring when you look at the pictures. Following this simple rule, I believe, could have prevented the incident thus making it the most important rule on the job. It was a simple wheel chock. Yes, the wheel chock.

A wheel chock is a tool used to keep the truck from rolling. It does not protect against electrical contact. The rule book is clear, "if the boom is in the air, the wheel chocks must be set." And I believe that this simple tool, this simple rule, much like making one's bed, could have triggered something greater. Thus, making it the single most important safety rule on the entire job!

In April 1987 Alcoa hired a new CEO, his name was Paul O'Niell. O'Niell was not an industry insider, actually the opposite. To this point, the 51-year-old executive had worked his career in government. Investors were nervous. To introduce O'Niell to the financial community Alcoa held a meeting in New York. Investors just wanted to see this new CEO and hear him say what all CEOs say; that they will maximize profits and see growth through synergies, rightsizing, accountability, and brand strengthening . . . you get the point. So, O'Niell stepped to the microphone. He greeted the investor community. Then, O'Niell outlined his big corporate strategy—safety. You could have heard a pin drop. If cell phones and texting had been mainstream then, investors would have been texting the word 'sell' back to the home office.

"I want to talk to you about worker safety," O'Niell began. "Every year numerous Alcoa workers are injured so badly that they miss a day of work. Our safety record is better than the general American work force, especially considering that our employees work with metals that are 1500 degrees and machines that can rip a man's arm off, but it's not good enough. I intend to make Alcoa the safest company in America. I intend to go for zero injuries." He went on to tell these investors, "In the unlikely event of a fire exits are . . . ." Investors were not impressed.

But, O'Niell was not daunted; he went to work. He faced the same question business leaders are faced with everyday, how can the hundreds, if not thousands, of employee habits be changed so that no one gets hurt? What O'Niell understood was that you start small. You start with one change as long as it is the right change.

Making the Bed Should Be a Safety Rule

If O'Niell had been in the utility industry he might have envisioned a wheel chock.

If I could start disrupting the habits around one thing," O'Niell said, "It would spread throughout the entire company."

Charles Duhigg, in his outstanding book, *The Power of Habit, Why We Do What We Do in Life and Business,* writes, "O'Niell believed that some habits have the power to start a chain reaction, changing other habits as they move through an organization. Some habits, in other words, matter more in remaking businesses and lives. These are keystone habits and they can influence how people work, live, spend and communicate. Keystone habits start a process that over time transforms everything. The habits that matter most are the ones that when they start to shift dislodge and remake other patterns."

Here is where bed making comes in. This one small habit, research finds, can boost productivity and happiness. Even though it's a simple bed, and no one else will ever know if you made it or not, it triggers something bigger. It's a keystone habit. The same can be said for wheel chocks.

When the boom hit the electric line killing one worker, the wheel chock would not have saved his life. But, setting the wheel chock is a basic rule. It is also the first thing, the first rule, that should be done after parking the truck—in other words all other safety rules and procedures come after setting the chock. The simple act of setting the wheel chock can be that one habit, a trigger, that signals to the entire crew that all rules will be followed. The fact that it was not set offers insights into the crew's frame of mind. No wheel chock means no formal trigger to cue other rules like a formal job briefing (plan), and safety rule compliance.

A wheel chock can be a keystone habit if it is both a rule (which it is) and a trigger of professionalism signally to the crew that it's time to huddle, discuss hazards and follow all rules. Under O'Niell's leadership, focusing on safety and a small change (keystone habit) Alcoa went on a decade of prosperity. It's profits hit a record high, net income was five times larger than before O'Niell. And its market cap grew to $27 billion. All of this and as Alcoa became one of the safest companies in the world—the company's injury rate fell to 1/20 of the U.S. average.

What is your keystone habit? And, did you make your bed today?

Source:
Duhigg, Charles, *The Power of Habit, Why We Do What We Do in Life and Business*, Random House, 2012.

2 The Safest Damn Utility in the Country

In June of 1997 Captain D. Michael Abrashoff boarded the USS Benfold, he was the new commanding officer. Benfold is a guided missile destroyer staffed with 310 sailors. This was Abrashoff's first sea command so he was undoubtedly anxious as he walked onto the ship. What he found did nothing to calm his nerves!

As was military tradition, the Navy rolls out the red carpet for a departing commander, and to welcome the new one. About two weeks before the command change routine work stops and the entire crew prepares the ship by painting it from top to bottom and preparing the deck for a dignitary filed reception honoring the departing skipper. If you have seen one of these arguably you have seen them all. At the ceremony, an admiral will give a speech about the great performance of the ship's departing commander. The event is supposed to be upbeat and full of energy. Then to overwhelming applause the retiring captain will stroll into the sunset.

On the afternoon of June 20, Captain Abrashoff boarded his new ship and watched the ceremony from a distance. The departing skipper looked happy to leave. He was flanked by his wife, children, and mother. After kind, but mostly untrue, words were spoken by an admiral, the public address system announced that the departing captain was walking from the ship and relinquishing command. The ship erupted with applause. "They were jeering, blatantly relieved to get rid of him," Abrashoff recalls. "I had never seen such open disrespect in my entire military career. I was stunned. I can still feel my face flushing with embarrassment."

The attitude toward the ship's departing captain was symbolic of his old fashion command and control style. The cold relationship between the leader and the roll players was also clearly reflected in Benfold's performance scores as it ranked near the bottom in most matrixes in the Pacific fleet. The command rotation is about 24 months, so Abrashoff knew that if Benfold was to improve, and his departing celebration was to be at least civil, he'd need to work fast.

Commit to Being the Best—For Captain Abrashoff, there didn't seem to be any middle ground to be a poor performing ship, or even a mediocre one, the stakes were simply too high. Every ship in the Navy's fleet was part of a team, and each team member had to perform at a very high level. Failing to do so didn't mean that the ship was ranked low in 'matrix scores.' Being a poor performer put lives at risk and in a small or large way jeopardized part of the United States' national security. For Abrashoff it was excellence or bust. To that end, early in his tenure Captain Abrashoff decided that Benfold would be the best damn ship in the Navy. Remember, to achieve this goal, he couldn't bring in new people. The same 'team' that cheered wildly when the departing Captain left were the same people that would need to take Benfold from last to first. Abrashoff said, "I decided Benfold was going to be the best damn ship in that Navy. I repeated it to my sailors all of the time, and eventually they believed it themselves."

The Path to Buy In

Once people begin to believe they are part of something great, like the best damn ship in the Navy, they begin to see things differently. For example, when performing a task, it is no longer okay to be just 'good enough' because the best damn ship does things excellently. Sailors begin to act differently when they are part of something greater than they did when they were just another sailor on just another ship. This mind set is very important. Hearing, them saying "you are best damn ship" is a start but leaders need to support that goal too. Here are some key steps to buy in:

Energy

Jim Loehr and Tony Schwartz, in their groundbreaking book entitled *The Power of Full Engagement: Managing Energy, Not Time, is the Key to High Performance and Personal Renewal,* simply state that, "Leaders are the stewards of organizational energy!" Abrashoff knew that his team couldn't get results without energy, enthusiasm and passion. He also knew that he was the one who set that tone. Abrashoff later wrote, "Leaders need to understand how profoundly they affect people, how their optimism and pessimism are equally infections, how directly they set the tone and spirit of everyone around them."

Show Up and Listen

Have you been talking to someone and after several minutes realized that he/she had not heard one word you said? He/she was not listening. Has someone been talking to you lately and you realize that you hadn't heard a word he said? Your mind was somewhere else. Have you been in a meeting lately and realized that no one was there! Everyone's mind was somewhere else. There is a simply concept called 'show up.' That is when you simply concentrate on the person or task immediately in front of you and eliminate distractions. When it comes to people, showing up and listening can have a profound effect on the trust level of you team. Abrashoff said, "Shortly after I took command of Benfold, I vowed to treat every encounter with every person on the ship as the most important thing at that moment."

Use Stainless Steel Bolts and Nuts

For over a hundred years sailors have spent a huge chunk of time painting the ship. Sea water can be very harsh on bolts, nuts and hardware. A thick layer of fresh paint can deter corrosion. But, even if you like to paint (which few sailors do) it is busy work that takes away time from important training and classroom modules—activities that make people better, and improve the ship's overall performance. One day, a 21-year-old sailor who had a paint brush in his hands a little too often had an idea. Why don't we buy nuts and bolts that don't rust, then we wouldn't have to paint so much. Abrashoff loved the idea. So, armed with the ships credit card, a small team searched for stainless steel replacements, and now the entire Navy uses them.

Respect, the Golden Rule

A sailor who had served the entire time for the previous Captain was stopped by that Captain one day and asked if he was a new sailor. Obviously insulted that his Captain had no idea who he was, he responded that yes, in fact he was new. The Captain proceeded to tell this sailor that he was the Captain, the one who was in charge. Abrashoff decided that he would meet with every sailor on his ship. It started with a one hour interview where Abrashoff was able to learn more about his crew, about their families, and individual hopes dreams and ambitions. "Our people don't care what we know until they know we care," said leadership guru John C. Maxwell. Abrashoff would simply say that this interaction supports being the best damn ship!

As Captain Abrashoff's time came to an end his commanding officer called and asked about the closing ceremony. Abrashoff said that he didn't want a 'big deal,' and that he had something else planned. Abrashoff, using monies saved from not buying paint and other efficiencies, ordered 310 live Maine lobsters and had them overnighted to Benfold. Over a nice lunch Abrashoff gave the shortest change of command speech in military history; it was five words. "You know how I feel," he said. And, with that Abrashoff relinquished his command. In his first year, USS Benfold went from last to first, winning the coveted Spokane Award, which is given to the top performing ship in the Pacific fleet. Or, as Abrashoff would say—the best damn ship in the Navy.

Source:
Loehr, Jim, *The Power of Full Engagement: Managing Energy, Not Time, is the Key to High Performance and Personal Renewal*, Free Press, 2003.

3 Safety Culture: What's under Your Helmet?

"In a weak culture, we veer away from doing the right thing in favor of doing the thing that's right for me."
—Simon Sinek

Jonathan Martin was a big athletic kid, and smart too. Martin's mother and father were both Harvard graduates. His mother worked as a corporate attorney while, his father was a professor at UCLA. Growing up in California, Martin attended some of the best schools in the area, including the Harvard Westlake School in Los Angeles. He loved his studies and football.

As a high school senior Martin was rated as one of the top fifty linemen recruits. To blend school and sports, he chose to play college football at Stanford. Martin studied Ancient History, and excelled at left tackle. As a redshirt freshman he made the freshman all-American team. Then as a sophomore and junior was recognized as a first team all-American. He declared for the NFL draft a year early and was projected as the third best offensive tackle. Martin was selected as the forty-second overall picking the 2012 NFL draft; selected by the Miami Dolphins. He was so excited for this next phase of his life.

Yet, sixteen months later Martin walked out of the Dolphins practice facility and shocked the football world by exposing a culture that was toxic. What does this have to do with worker safety? Everything.

On February 14, 2014, nearly four months after Martin walked out of the Dolphins facility, the NFL released a 144 page report on their investigation dubbed the 'Ted Wells Report' after lead investigator and well respected attorney Ted Wells. The study found and detailed a workplace culture that allowed bullying and harassment.

Wells writes, "After a thorough examination of the facts, we conclude that three starters on the Dolphins offensive line, Richie Incognito, John Jerry, and Mike Pouncey, engaged in a pattern of harassment directed at not only Martin, but also another young Dolphins offensive lineman, whom we refer to as Player A for confidentiality reasons, and a member of the training staff, whom we refer to as the Assistant Trainer. We find that the Assistant Trainer repeatedly was targeted with racial slurs and other racially derogatory language. Player A frequently was subjected to homophobic name-calling and improper physical touching. Martin was taunted on a persistent basis with sexually explicit remarks about his sister and his mother, and at times ridiculed with racial insults and other offensive comments."

The report documents countless vulgar statements and instances of harassment. One example from the report, "Incognito recorded a $200 fine against himself for 'breaking Jmart.'"

Safety Culture Is Under the Helmet

Culture has been simply defined as 'what is acceptable around here.' What was acceptable in the Dolphin's locker room was a sick climate of harassment and bullying. But, this is not much different than a culture that allows unsafe acts. One that encourages or keeps quiet about safety rule violations. If you swap the words 'harassment' with 'at-risk act' is it really that much different? There are several key take-a-ways for all leaders. Consider these points.

Culture Is Ground-up

"That ultimately rests on my shoulders, and I will be accountable moving forward for making sure that we emphasize a team-first culture of respect toward one another," said Dolphin's head coach Joe Philbin after the incident. The Wells report found that Philbin had no knowledge of what was going on in the locker room he was responsible for. Culture grows from the ground up. And, the way to really know what is going on—on the ground—is to be on the ground.

Safety Culture: What's under Your Helmet?

People Who Can Change It, Can Change It if They Speak Up

Former NFL football player Mark Schlereth in a blog straight from his heart entitled "Don't lose crucial parts of 'the code,'" told this story.

"In my seventh season, I found myself on a bus in Japan as a member of the Denver Broncos. It was my first season in Denver and our first road trip of the preseason. As we sat in traffic, there was the usual joking and poking fun that accompanies those moments. In the seats behind me sat two defensive players, and they were flipping some grief to a young player, typical stuff. At some point, the good-natured, innocuous ribbing became personal and out of bounds, so I turned and said 'Enough,' they responded with a few choice words for me and I made it clear in no uncertain terms that they crossed a line and I wasn't putting up with it. They mumbled a few protests under their breaths, but it was over and the bus rolled slowly to its destination, I glanced back at the young player I had stood up for—no words were exchanged, just a tacit nod of the head, as if to say, 'Thanks. I appreciate the help.' I replied in kind, and it's was never brought up again. They got a little carried away, but they knew I was right. We moved on with no trouble. Nothing lingered or simmered because it was addressed on the spot."

Does your safety speak up, or is more like the Dolphins? Only those in the culture on the ground can speak up to make change happen.

Culture Is under the Helmet, Not on the Wall

As a dad of a teenage girl and 12-year-old boy I remind myself daily of the saying that reads, "Don't watch what I say, watch what I do." In safety that is often translated to, "It's not what the posters on the wall say . . . it's what happens in the field." The Ted Wells report clearly documented that every player and coach involved had received and signed the Dolphin's work place harassment policy that strictly forbids any such action that happened. Sound similar to our safety rule book, do we follow those rules? To know for sure, one must get under the helmet (hard hat) and actually work a day in the shoes of your workers. Do your posters align with what is happening in the field?

Mark Schlereth finished his blog by writing, "I'm left with this conclusion about the Dolphins organization from the coaching staff on down: They were either complicit, incompetent or, worse, both." What would be written about you, your staff and your culture if a fatality happened in your company today?

4 Scoring 100% Safe Work

A Leadership Secret to Success in the Utility Industry

Can You Answer These Questions?

▸ A bat and a ball cost $1.10 in total. The bat costs $1.00 more than the ball. How much does the ball cost?

▸ Next, if it takes five machines five minutes to make five widgets, how long would it take 100 machines to make 100 widgets?

▸ Finally, in a lake, there is a patch of lily pads. Every day, the patch doubles in size. If it takes 48 days for the patch to cover the entire lake, how long would it take for the patch to cover half of the lake?

Congratulations, you just completed the *Cognitive Reflection Test,* or CRT. The CRT is the brainchild of Yale professor Shane Frederick. And, the purpose of the test is to measure ones cognitive abilities in a simply, fast and fun way versus the traditional tests that come with hundreds of questions and take hours to complete.

To prove his point, that this is a 'worthy' test for cognitive ability, Frederick gave his test to several college students across a diverse set of college campuses.

"Frederick gave the CRT to students at nine American colleges and the results track pretty closely with how students from those colleges would rank on more traditional intelligence tests. Students from Massachusetts Institute of Technology, perhaps the brainiest college in the world, averaged 2.18 correct answers out of three. Students from Carnegie Mellon University in Pittsburgh, another extraordinarily elite institution scored 1.51 right answers out of three. Harvard students scored 1.43, the University of Michigan at Ann Arbor scored 1.18, and University of Toledo 0.57.

If you wanted to improve test scores, what do you think would be the best option? Make the test easier, right? Actually, if you wanted to improve test results you would actually make the test a little more difficult. Princeton psychologist Adam Alter and Daniel Oppenheimer did just that, they made the CRT just a little more difficult by printing it in a hard to read font. Those taking the test had to read then reread the questions. It was this small but meaningful change that improved results. For example scores for students at Princeton went from 1.9, on average for three questions, to 2.45.

Some of the most difficult, complex and unusual work in the utility industry is during storm recovery. Whether it is in the aftermath of a hurricane, a tornado or just a typical thunderstorm, utility work is the most difficult and hazardous in these situations. And, it is traditionally the safest in terms of fewest injuries. Why? The fact that it is a little more difficult does several things for utility workers. Workers plan better. They communicate more effectively and overall safety awareness is much higher. As Alter and Oppenheimer said, "Suddenly you have to work to read the question . . . think more deeply about whatever they come across and use more resources on it, they will process more deeply and think more carefully."

By contrast, we also know that some of the most simple and mundane utility jobs have led to catastrophic injuries and even fatalities. Why? Just like these three seemingly simple questions, we don't stop to think. What Alter and Daniel Oppenheimer discovered is that putting a simple step in place, the difficult font in the case of the CRT, to make people think for just a second longer than they would otherwise, improves results—the same is true in utility safety.

By the way, the right answers are: 1) Five cents (not ten), 2) Five minutes (not 100), 3) 47 days (not 24) and put something in place today to make all utility workers think just a little longer about the work at hand. If you do they will score 100 percent safe work!

Source:
Gladwell, Malcolm, *David and Goliath: Underdogs, Misfits, and the Art of Battling Giants*, Little, Brown and Company, 2013.

5 How to Slow Down and Avoid Injuries on the Job

Every lineman has a time when he or she has rushed, bleary-eyed and frustrated through a job. By rushing through a task, however, linemen can put themselves and their crew in danger.

I learned my lesson several years ago when several hundred homeowners were out of power. I rushed through a job, only to spark a fire at a substation.

It all began when I worked a trouble call on my own. As a new journeyman, it felt empowering to check meters and services, clear limbs and refuse cutouts. In the middle of one order, however, the dispatcher called. The 34.5-kV circuit that fed three small towns had locked open. He told me to immediately go to the substation for switching instructions. He advised me to open the breaker and associated switches so we could do some line repair and then close the switches and breaker to restore power to more than 1,000 customers.

I had a total of three instructions, and I accomplished the first two without incident. The final order, which involved opening a certain switch, became a problem, because I could not find the switch number. I was irritated and frustrated that I could not find the switch location. When I finally found it, I couldn't find a switch handle. I shined my flashlight through the heavy rain and substation steel to see the switch. I was in a feverish pace at this point, and I looked everywhere for a switch handle so I could operate the switch.

After what seemed like an eternity, I found a handle, inserted it into the switch and operated the switch. It was about halfway through that procedure when I knew something was wrong. As I finished the operation, I immediately knew that I had closed the switch, which was exactly what I was not supposed to do. Because I was hurrying and annoyed, I had just closed the bypass switch, re-energizing the line. Without thinking, I immediately threw the switch open. A huge fire ensued, because I dropped the load of three small towns with a solid blade switch.

In the end, relays at a bulk substation put the fire out. No one was hurt, and I was extremely lucky that no equipment was damaged. What I didn't know then, but understand now, is that I broke two of the H.U.R.T. (Hurrying, Upset, Rerun, Tired) laws. If we can recognize when we are in H.U.R.T. mode, we can go a long way to preventing incidents and injuries.

Hurrying—Legendary college basketball coach John Wooden told his players, "Be quick but don't hurry." Make sure that you do not lose sight of the big picture and make costly mistakes. If we find ourselves hurrying on a job, we must take a two-minute "safety stop." During this "time-out," re-evaluate all surroundings, hazards, and attitudes before proceeding with a task.

Upset—A few years ago when I was working as a safety professional for a Midwestern utility, I received a call that a lineman had contacted 12 kV. I immediately drove to the location and interviewed all involved. The crew had energized a section of line and then took lunch. Immediately after lunch, one of the crew members went back up in the air to make jumpers, completely forgetting the line was energized. He was lucky that he didn't get seriously hurt. What we discovered later on was that during lunch, he took a call from home and was upset. When you're bothered, frustrated, irritated, annoyed, angry or discouraged, discontinue work and perform a safety stop.

Rerun—Have you ever taken an hour-long drive and found yourself zoning out and thinking of other things while you were behind the wheel? The same thing happens when we are working on a routine task. When we know a task is a rerun, we can find ourselves losing focus. Complacency kills, and you must perform a safety stop to get your mind back on track.

Tired—The disastrous crash of the Challenger space shuttle in 1986, in which seven astronauts lost their lives, occurred after NASA officials made an ill-fated judgment call to go ahead with the launch after working for more than 20 consecutive hours, according to Jim Loehr and Tony Schwartz in the book *The Power of Full Engagement*

According to the authors, the longer, more continuously and later at night you work, the less efficient and more mistake-prone you become. In other words, when we're tired, it's like we're driving on ice. We know we have to move ahead, but we do so with much more caution. If you find yourself working tired for any number of reasons, perform a safety stop, then move forward.

Once you recognize a H.U.R.T. sign, perform a safety stop immediately. Identify the source of the H.U.R.T., and form a simple plan so that work may continue safely. Once the H.U.R.T. is gone, the chances of injury or incident decline.

6 Leading with Chaos . . . and It Is All Chaos!

Four Steps to Help You Lead in Times of Confusion, Pressure, and Disorder

Leading with chaos is not about using confusion and turmoil in an attempt to get results. Instead, it is a simple recognition that often work life is chaotic. Recognizing chaos, the challenges and opportunities it brings and then leading alongside it, will help both your career and your team get results! Let me explain.

I just finished a three month long project and my life while working on this project can best be described as 'chaos!' Weeks easily stretched from sixty to seventy to eighty hours or more. Conference calls beginning at 7 a.m. on Sunday mornings. Saturday afternoons spent preparing an unexpected memo for a senior manager or outlining key objectives for the next week. Excusing myself and stepping out of a family function, a movie with the kids or even a special night out with my wife to respond to a an urgent phone call or email. The pace was frantic, demanding and exhausting and the dynamics often perplexing and disorderly.

Chances are, in the last year or two a work related project has engrossed you too. When it does, life within this project can best be described as 'chaos.' Whether you were on a team working feverishly against a drop-dead date to publish the annual shareholder report, a utility company working twenty hour days per day for weeks on end to restore power to hundreds of thousands of people after a major storm or power plant outage, hustling to meet a deadline to put a new product on line, working to get a piece of vital equipment back in service or a project critical to you and your company's success the dynamics within these projects are generally the same.

In addition to endless hours of work, emails, conference calls and meetings at all hours of the day and night, there are other dynamics that are difficult to get ones arms around. Often, in these settings, there are multiple teams, vendors, consultants and companies all working together to meet objectives. These teams come with different cultures, norms and communication styles and time zones. There are organization structures, charts and multiple leaders all of whom need to be updated and informed. There are rank and file teammates who all have different bosses, objectives, compensation toward the goal and alliances. Within these teams there can be competing goals and timelines, where one team is willing to sacrifice your goal to meet theirs. Chaos, by definition, is said to be a state of disorder and confusion. It is further described with words like 'unbounded' or 'formless.' Chaos seems to be a perfect word to describe these work projects.

I might also note that with today's life style of smart phones, world markets, 24/7 news cycles and business demands, the thin line between work and home life, some believe it is all chaos!

Chaos in our work life and within projects is here to stay. Yet another interesting dynamic of chaos is that leadership is both more important for results and at the same time more absent than ever. In chaos, teams put up walls, silos, and both individual and team energy drains. When this happens people will naturally become less flexible, their communication circle shrinks, and they are generally less tolerant and understanding. Leadership across the entire project defaults to a 'check the box' attitude with individual teams sharing less information. Chaos needs leadership! In chaos there is opportunity for you to lead right beside it. In so doing you will better support your team, the project and your personal growth and development. Here are five thoughts on leading in chaos.

Remember Why!

There is a legendary story involving Herb Kellerher, the founder of the Southwest Airlines Company (SW), that can help us remember why we do what we do. The story starts with one of Kellerher's senior vice presidents bursting into his office. The senior VP immediately explains to Kellerher that there is a major problem.

At issue is the fact that a number of competing airlines are offering a light dinner on the Los Angeles to Las Vegas flight and projections are indicating this trend might cause SW to lose market share to these other companies. The senior VP was suggesting that SW immediately add the same feature to their flights. The story goes that Kellerher listened intently then responded with something like this, "Does adding this meal make us the lowest cost airline?" The answer was simply no. So, no meal added, problem solved, end of discussion.

The point is that you and your team are not the project, and you are not the chaos. You work to provide something greater—that is the 'why.' SW strives to provide the lowest cost. Medical providers save lives. Utilities provide life and light. Financial institutions provide college to teens through savings programs. In the midst of chaos however, the why is often forgotten. A dynamic of chaos is that small things, like a meal provided by a competing company, become big deals. Leading in chaos means continuous focus on the bigger picture—the why.

Organize Blue Chips!

Often, if I am conducting a full day or multi day seminar, I will end the session with the blue chip activity. Two participants will stand with their backs to a table as the rest of the participants gather around to cheer on these two volunteers. When I say the word, they will turn and begin to pick up discs. After about ten seconds I will yell stop. What generally happens is that the participants will turn around and immediate start picking up chips, the ones that are closest to them. What they fail to realize is that there are three different color chips, white, red, and blue. When they turn around, the white chips are right in front of them and there are lots of white chips, so they focus their entire time picking out these chips. What they don't realize is that white chips have a value of one, red chips have a value of ten, while blue chips have a value of 1,000! In a recent session, I asked a participant why he didn't pick up blue chips, he said, "I didn't even see blue chips!" He was totally focused just on those chips immediately in front of him.

The game is a solid reflection of life in chaos. After days or weeks of 'chaos' we get in the habit of simply getting things done, and done quickly (grabbing chips) without stopping to think of the value of that activity. In chaos the workload can be overwhelming, but not all tasks bring the same value to the project. Spend time each day, and then before the start of each week, to outline blue chips—activities that bring great value. This could be a simple coaching session with a direct report, a key meeting, or interaction with the CEO. Whatever the blue chip, capture it and make sure it gets completed. Leading with blue chips means the important things are getting done and energy is spent where it is most valuable.

Live Above the Line

The dynamics of chaos are interesting. In chaos, silos become stronger, not weaker. Sharing and volunteering happens less often, not more. People are actually less flexible, and less understanding. Because of these dynamics, and others associated with chaos, people can often lean on emotion to enforce their point, not fact. And, people will more frequently dip below the line. It is important for you as a leader to always stay above the line. Below the line is personal attack; leading with emotion, not fact; masterminding and second guessing decisions after the fact; hanging in a clique, and not sharing outside of the circle. What living above the line means is that you are professional. You lead with fact and do not let emotions carry your tone, attitude, or response.

Energy

Jim Loehr and Tony Schwartz, in their insightful book, *The Power of Full Engagement: Managing Energy, Not Time, is the Key to High Performance and Personal Renewal,* addressed the importance of energy. They wrote, "Energy, not time, is the fundamental currency of high performance." They also said, "Leaders are the stewards of organizational energy!"

Chaos will drain both your team's energy and your energy. When energy is low, your quality of thinking is lower, your communication is less effective and you have less of an ability to understand other opinions. When working in chaos be mindful of the importance of energy and continually gauge both your personal energy and your team's energy. Find ways to infuse your team with energy, and have strategies to keep your energy high too.

In closing, remember, "When nothing is sure, everything is possible" (Margaret Drabble). In today's work world, you will find yourself working with chaos sooner not later. The core question is, will you be in a position to lead with chaos. Your team, and your career, need you to be ready.

Source:
Loehr, Jim, *The Power of Full Engagement: Managing Energy, Not Time, is the Key to High Performance and Personal Renewal*, Free Press, 2003.

7 Thinking Differently about Target Zero

Over the last decade many companies and utilities centered the goal of safety to target zero. It is both a journey (to achieve zero injuries and incidents) and a destination (a milestone to be achieved each day, week, year over year). But now that some companies are more than a decade into target zero, it might be time to think about it a little differently.

On a seemingly unrelated note, last year I was reacquainted with integers—do you recall integers from your junior high math class? My daughter was studying the concept and needed some help. From kindergarten to junior high, math numbers range on the 'positive' side of zero. In preschool for example, numbers were simply 1, 2, 3, . . . up to 10. Later we learned numbers went from 10 to 20, then to 100, then a thousand—you get the picture. Our calculations (adding and subtracting then multiplying and dividing) were all in the ranges just noted above, zero or greater. But, as kids get older and more mature, at least in their thinking of math and numbers, there is a new concept introduced (integers). This is the reality that numbers don't stop at zero, the number line actually goes past zero, to negative 1, negative 2, negative one million, and beyond. In fact, when looking at integers, zero is just the center point on a number line and there are an equal number of 'numbers' on each side of zero.

In the late 1800s less than a dozen men met in St. Louis, Missouri. They were electrical lineman in this new and emerging industry. In the meeting they decided to do something radical. At that point in time, safety was 'trial and error.' On average, one out of two men who began a career as a lineman would be killed on the job. This group of men formed what would eventually be called the International Brotherhood of Electrical Workers. They wanted better working conditions, pay—and safety.

From those roots, safety for line workers and utility workers began to improve. Standards developed, as did working clearances, national electric safety codes, PPE, then OSHA. The industry went from the bleak fact that nearly one in two workers were killed on the job, to a much more sustainable safety program. Yet, one hundred years later, the industry looked within and didn't like what it saw. While fatalities were infrequent, men and women were getting hurt. Total injuries still numbered in the dozens or hundreds. And, serious life changing events were still occurring, and all too frequent. One by one, many utilities began to shift values, moving to include safety as a value and live that value through a theme, 'target zero.' Today, numbers are much lower with some utilities going long stretches with zero incidents or injuries.

Target zero is absolutely the right safety value, but moving forward our thinking of target zero will change. When it was first introduced over a decade ago, it was clearly a destination, an aspirational goal. Yet, today, with the successes and improvements we have had in eliminating injuries and improving work conditions, we should starting to slowly understand, just as my daughter did in her math class, that zero is not an end point a stopping point or a destination, but a mid point. In safety we can clearly hit target zero for injuries and also employ safety programs and strategies that take us far beyond zero. Don't believe me? Well, that is the first place to start!

Believe—Henry Ford coined the following phrase, "Whether you believe you can or believe you can't you are generally correct." If you would have asked line workers in 1900 what was possible in utility safety they would have never said target zero. They couldn't fathom working conditions where workers and management alike shared that value. But, about a decade ago beliefs began to change, and target zero beliefs were born. Today, we are just starting to push thoughts and programs that actually send our workers home in better condition in which they came. These programs include health and wellness programs. Motivational and leadership programs that tap into our workers talents. They provide our workers training and tools that give them more confidence as they walk out the door at night then they had when they walked in that morning. I believe we are ready to move beyond zero and not only provide a work place of zero at risk acts, but begin to move beyond zero. What do you believe?

Expect It—Today, many target zero posters and value statements include something about 'being responsible and accountable for my safety.' That is a great and necessary part of the target zero process but over the last decade it has all too many times been understood by our workers to mean 'discipline' for breaking safety rules. While employers must be proactive in rewarding positive actions and redirecting choices that are not aligned with target zero. As we move beyond zero safety accountability begins to take on an expanded meaning, "What more can I do for safety results."

Living Safety—A number of years ago Bruce Larson went to a corporate board foundation. The foundation funded research and Bruce had an idea. Bruce wanted to tour the United States and Europe and interview the most successful business leaders and politicians of the time asking them one question—If you had to sum up success in one word what would that word be?

The Foundation liked the proposal and funded the research. After two years and hundreds of interviews, Bruce returned to inform the board he had found the secret. The one word secret to success—risk.

But, Bruce understood that 'risk' didn't mean taking a chance or a shortcut, he categorized risk into a number of areas, one being emotional risk. Emotional risk is when you do something you are a little nervous to do, it is positive and powerful and it is for yourself or for someone else.

If we are going to hit target zero, then move beyond zero, we need to take emotional risks each and every day. Living safety means you take emotional risks. We give feedback to our coworkers. We stop jobs to review hazards. We ask about job planning and we check for all PPE and rule compliance. Living safety is giving and taking emotional risks (feedback).

Target zero is today's value. In time the industry will move beyond zero and values will shift. Believe, Expect, and Live!

8 Unreasonableness: How Leaders Make Progress

"Innovators need to be disagreeable," Malcolm Gladwell writes in his eye opening book called *David and Goliath: Underdogs, Misfits, and the Art of Battling Giants*, "by disagreeable that doesn't mean obnoxious or unpleasant... Social Risks to do things that others do not agree with. This is not easy. Society frowns on disagreeableness. As human beings, we are hard wired to seek the approval of those around us." Think about this story.

I can only imagine what Vivek Ranadivé was thinking when he walked into his first basketball practice. Despite his usual calm demeanor he had to have some terror rumbling in his belly, After all, Ranadivé was coaching 12-year-old girls. He only volunteered because as a single dad, and a very busy professional (CEO of his company), he looked for any and all opportunities to spend time with his daughter Anjali.

Vivek Ranadivé would have probably summed up his knowledge of basketball in a few words; never played before. Ranadivé grew up on India. There he played cricket and soccer, not basketball. I can almost feel the nerves in Ranadive's stomach. I'd wager that more than once he second-guessed his choice to coach.

It is true that Vivek Ranadivé did not grow up with basketball nor did he play it. But, it is also true to say that Vivek Ranadivé is successful, and that he might also find a way to make a 12-year-old junior national basketball team successful too. Vivek Ranadivé grew up in Mumbai. When in high school he heard of M.I.T., and decided that is where he wanted to go to school. This was in the 1970s and at that time, one had to have the permission of the Indian government to go to school abroad. This meant that one had to have a form signed by a certain government office. Vivek Ranadivé camped outside those government offices until he finally secured permission. Ranadivé did well in his studies and focused his career in computer software. Today, Vivek Ranadivé's leads Tibco. Tibco is a company founded by Ranadivé and it has digitized Wall Street, so the brokers have real time information. In fact, Tibco processes and transfers more data in one day than Twitter does in an entire month!

Vivek Ranadivé sized up his basketball team, and his talent. These were mostly girls who wanted to be with their friends, not series basketball players. With the exception of a few girls, most had not played before, or if they had it was a very limited recreational league. Ranadivé quickly adopted two coaching strategies. First, he would never raise his voice. He thought that using the same tone and example the he uses to lead his business would be more effective with 12-year-old girls than ranting and yelling. Second, he would spend the vast majority of practice time working on defense, not offense—Vivek Ranadivé's team would play the full-court press.

For Ranadivé it was more a math equation than it was a brilliant coaching move. With few exceptions, his girls could not shoot the ball. Learning an offense, where all five players work in unison with offensive skills of movement, picks, passing, cutting and shooting seemed like a longer term endeavor. And, the basketball court was 94 feet long. Traditionally, teams would retreat to play defense, yielding 70 feet of the court to the other team, and only playing defense in about 20 feet of space. Good teams, the kind that Vivek Ranadivé's girls would be playing, could then use their talent, offensive skills and strengths to impose their will. Ranadivé focused (and practiced) for two key timelines. The first is the five seconds a team has to throw the ball in bounds. The second is the ten seconds that a team has to break half court. If his team could steal the ball, mathematically they would have more possessions, and more easy lay ups than the team they were playing. On paper Ranadivé thought it would work.

And, it did work! Vivek Ranadivé coached his daughter on the Redwood City team. They won their league regionals, and they won two games at the junior nationals before finally being beat. And they won their games doing one thing—playing stifling full-court defense. George Bernard Shaw wrote, "The reasonable man adapts himself to the world. The unreasonable one persists in trying to adapt the world to him." Ranadivé was able to adapt the basketball court to his world (through his team's full court defense). But, as you can imagine, not everyone was happy with this style of play. Gladwell included some feedback, "The trouble for Redwood City started early in the regular season. The opposing coaches began to get angry. There was a sense that Redwood City wasn't playing fair—that it wasn't right to use the full-court press against 12-year-old girls, who were just beginning to grasp the rudiments of the game. The point of basketball, the dissenting chorus said, was to learn basketball skills. Of course, you could as easily argue that in playing the press a 12-year-old girl learned something much more valuable—that effort can trump ability and that conventions are made to be challenged.

Innovators, those who see the world differently and act on those insights, are not always met with open arms. Today utility safety lacks innovation. Without innovation, it means the safety results we get today will be, more or less, the results we achieve in five years . . . in ten years. Sure, we can make incremental improvements, and our injury numbers can trend the right way, but our results will lack the 'hockey stick' change that innovation can provide. We can learn much from Vivek Ranadivé, his 12-year-old junior national team, and seeing the world differently. Here are some places to start.

Find That One Thing—Vivek Ranadivé focused on one thing that could make the largest difference to the outcome (defense and the full court pressure). What is the one thing that can make the biggest difference in our worker's safety?

Welcome Outsiders—Next, Ranadivé didn't have 'the curse of knowledge' when it came to basketball coaching. Traditional coaching of youth says to teach fundamentals (dribbling, shooting, and passing) and over the years players will develop. But, since Vivek didn't grow up playing basketball, he didn't know that. Find people working for your utility that can observe your safety program and give opinions, ask questions, and suggest strategy. You can't see what you are blinded to—like Vivek Ranadivé, these coworkers of yours will see the world much differently.

Don't Worry about Critics—worry about results. A few years ago someone asked a very notable St. Louis Cardinal baseball player if the teammates got along. This was a player from the 1980s when the St. Louis Cardinals had World Series caliber teams. The player said, we didn't all get along but you know the best medicine for poor relationships? Winning. Vivek Ranadivé's team won, which fueled his players even more to buy into defense. Find wins for your team—and keep winning!

Throw It All Out—and justify what comes back. There is a senior leader of a corporate diversity team who meets with her advisory council every other year. They throw out all of their programs and start with a clean white board. This keeps them from being 'stuck' with good programs that could be replaced with great programs. This is something your safety program and process could adopt.

Use Science—Vivek Ranadivé used the insights gained from running a very math based IT company to give his team an advantage (more possessions and easier shots). What opportunities do you have to use 'science' to give your team an advantage? Use incident data, previous injuries, worker feedback, and any research at your disposal to find your advantage, and once you find it teach it to your workers.

95/5—In truth, Vivek Ranadivé's practices focused 95 percent of their time on defense. They practiced the one thing that they were good at, and they were good at it because they used their practice time to get better. How can you use your safety meeting and training time to really focus on the one or two things that can make 95 percent of the difference.

In the end, I agree that Vivek Ranadivé taught the girls more through this experience than a traditional dribble, shot and pass coaching technique. What can you learn, and apply, to your team?

Source:
Gladwell, Malcolm, *David and Goliath: Underdogs, Misfits, and the Art of Battling Giants*, Little, Brown and Company, New York, 2013.

9 Macho Safety
The New Definition of Toughness

The second Saturday of September 2012 held the long awaited conference opener for the Missouri Tigers football team. The Tigers, who transferred from the Big 12 to the South Eastern Conference, or SEC, were playing their first SEC conference game. Some consider the SEC as the strongest football conference in the country, and their teams have backed it up by winning the last six football national titles. The Tigers conference opener was a nationally televised game in Columbia, Missouri, against a top-ten ranked and perennial power, the Georgia Bulldogs. The game began as a defensive struggle and remained close. Mizzou, led by their junior quarterback James Franklin, was leading in the third quarter, 17 to 9, but mistakes by MU and strong defense and tackling by Georgia led Georgia to a win. The next team on the Tiger's schedule was Arizona State.

The Friday before the Arizona State game, rumors began to surface that James Franklin, the Tiger's starting QB, would not start against Arizona State. While the University of Missouri Sports Information Director was silent on the issue, sports reporters and bloggers were buzzing for facts. The rumor said that Franklin, who was sacked numerous times against Georgia, had a serious contusion on his throwing shoulder. This was a big deal for a number of reasons. First, the Tigers had not been without their starting quarterback for ten years—for nearly a decade their starter had not missed a start due to an injury. The backup to Franklin was a redshirt freshman with absolutely zero experience. And, Mizzou's game against Arizona State was a must-win. No one was saying whether Franklin would play, or if the rumors were true.

Moments before kickoff, head coach Gary Pinkel confirmed the rumors when an ESPN sideline reporter caught up with him. When asked if Franklin would play, Pinkel replied, "It was just too painful for him, and he didn't want to play." Pinkel went on to say that Franklin refused a painkilling shot.

These comments did nothing to silence the 'buzz,' instead they fueled the fire. Pinkel, who has been Mizzou's football coach for a dozen years and is the winningest football coach in the school's history, is predictable. Any sports reporter who has followed Pinkel could have forecasted Pinkel's comments concerning Franklin. It would have read, "Franklin was evaluated by medical staff and in their opinion, he is not able to play. They make all of those decisions, not me."

But that's not what Pinkel said. He said Franklin "Didn't want to play," and refused a painkilling shot (as if taking that shot is the 'norm'). Reading between the lines, Pinkel's statements could be interpreted to say, "From one football player to another, you aren't tough enough to play."

After the game, Franklin told the *Columbia Daily Tribune,* "I was just telling them I didn't think it would be smart to go in. I just said I wasn't 100 percent confident (in my shoulder). I like to feel the pain to be my own judge of it," Franklin continued. "I don't want to numb it and then play and possibly hurt it more. The next thing I know I won't be able to play for a couple weeks."

What does 'toughness' really mean? This story gives many insights into the macho attitude surrounding college football. Players get hurt. When they do, players are expected to take painkilling drugs in order to play. If you refuse, you are called out, "He didn't want to play." But what about the attitude that says I don't do that. Sure, I play football, but I have a different mindset toward dealing with pain and injury. An individual who thinks, "These tough guy norms don't apply to me . . . I don't take painkilling drugs. I understand my body and my limits. I want to help the team today but also long-term, and I want my body to be around after college." That mindset is macho too—and maybe tougher.

We've all been around the utility business long enough to understand that our industry has a macho attitude. An attitude that says, I know what is best, or this won't hurt me or, that rule doesn't apply to me, or just because the rules say to do it that way, we do it this way, or finally, training says to do it that way, but that is only in the classroom, we do it this way in the field. In short, over a thirty year career I'm going to get hurt a time or two, because that is just what we do—this is the macho attitude that has been around for decades. But, there is a new attitude emerging that says we don't have to take shortcuts. We don't need the painkilling shot. We don't have to get hurt at work. This is the new macho safety, and to get there think about these three steps.

Follow Coach Mike—Duke basketball coach Mike Krzyzewski is likely the most effective coach in any current sport at any level of competition. His success is unparalleled. That said, he coaches and mentors 18- to 22-year-old young men who have all of the pressures of being 'rock stars.' His players are watched, televised, analyzed, interviewed and reported on—all by the national media. In order to maintain control of his players, one would think that Coach K has an entire book of rules to protect and guide his players. How many rules do you think his program has? One. That rules reads, "Don't do something detrimental to yourself." Coach Krzyzewski believes that this one principle is all that is needed. In taking care of yourself, you also take care of your teammates, coaches, and school. I think that James Franklin knows this rule, and each of us and our workers should too.

Think Long-Term—In September 2012, the U.S. Bureau of Labor Statistics (BLS) released their annual data on fatal occupational injuries for the previous year. In 2011, the report shows that in the U.S. 4,609 people died from on-the-job injuries.

American Society of Safety Engineers (ASSE) President Richard A. Pollock, CSP, said people should be concerned. "It is alarming that 13 people a day are dying from work related injuries. This is a serious problem that we find unacceptable," Pollock said of the report. "These incidents can be prevented. We urge all companies and organizations to take measures now to make sure they have developed and implemented management systems of control that include effective occupational safety and health programs aimed at preventing worker fatalities, injuries and illnesses.

I agree with Mr. Pollock that we should be concerned and that our industry needs to continue to develop systems to ensure safety. That said, I also believe that safety needs more people like James Franklin. James represents an attitude that says "I don't take risks... I put myself first." Our business all too often has an attitude that condones shortcuts or cut a corner in order 'git ur dun!' If each of us can have that new macho safety attitude, 13 people won't die on the job each day!

Plan Before You Play—Before kickoff, a college team has an entire week of practice and preparation. In addition to practice, there is film review, team meetings, weights and conditioning, a game plan, a practice squad, more practice, more film review and revisions to the game plan. Each player understands his role and each coach is specific about the job that needs to be done. That said, if football players can prepare for an entire week, we can take the time needed to plan our work and make sure that every crew member understands their job and the hazards associated with it. Plan before we play—yes that is the macho thing to do.

Each Monday the Missouri football team hosts media day. Coach Pinkel and James Franklin put this issue to rest and the team turned its focus to their next game—a road trip to 7th ranked South Carolina. Having rested his shoulder James is expected to be the starting quarterback. What is macho football (and safety), well, that depends on your definition. Work safe!

Footnote: the year following this season, when Franklin was a senior, he led his team to a SEC East championship and a 12–2 record... now that's macho!

10 What Toothpaste and Safety Leadership Have in Common

Can you feel it? If you run your tongue over your teeth you will feel a film. That film is what Pepsodent (in the 1920s) was designed to remove. And, the result of using Pepsodent would make your teeth beautifully white in the process.

It's hard to believe now, but do you know that in World War I, the United States military listed poor dental hygiene as a national security risk? In the early 20th century as the country's wealth grew, United States citizens were eating more processed foods and foods with much more sugar. Couple this with the fact that there were no toothpaste brands on the national market. Toothpastes at the time were still sold locally or regionally, mostly by traveling 'door to door' salesmen. These 'so called' toothpastes were a concoction of all sorts of things.

Claude C. Hopkins, was on the top of his game. Hopkins was riding the very top of a new wave—advertising. He, and his firm, had made brands like Van Camp's pork 'n beans, Schlitz beer, Palmolive dish soap, Quaker Oats, Bissel Carpet Sweeper, and Goodyear tires household names. Hopkins was doing so well that in his autobiography he complained about how difficult it was to spend all of his money!

As Hopkins worked one day an old friend dropped in. This friend had a new product—a toothpaste called Pepsodent. The investor group was a little 'suspect.' It was rumored that one investor had some land deals that had gone south. Another investor was likely in the mob. Hopkins thought that an ad campaign for toothpaste was the equivalent to professional suicide. American's were happy with the door-to-door sales, and besides no one brushed their teeth anyway. He politely declined.

Hopkins friend was tenacious however, returning time and again until finally Hopkins made an outlandish offer. For a large stock option in Pepsodent, Hopkins would write an ad campaign. The two shook hands.

What Toothpaste and Safety Leadership Have in Common

Begrudgingly Hopkins went to work. He began reading dental textbooks to understand more about dental hygiene. I'm sure he thought the material was the equivalent of a 'root canal.' But, late one afternoon he discovered that teeth have something called mucin plaques. Hopkins re-read the material because the concept of mucin plaques, Hopkins thought, might be the key to selling Pepsodent. Hopkins ran with it, and decided to call the mucin plaques 'the film.' He designed an ad campaign around this 'film.' One such ad read, "Just run your tongue across your teeth, you'll feel a film. That's what makes your teeth look off color and invites decay." Did this approach work? Very well!

Agreeing to run Pepsodent's ad campaign for stock options proved to be the most fruitful financial deal of Hopkins' career. Within five years of the launch of the ad campaign, Pepsodent was one of the best-known products on earth. In the process, the product established a tooth brushing habit across the United States that may have 'saved' the teeth of millions of people. By 1930 Pepsodent was sold in a number of countries. A decade after the campaign launch, tooth brushing was a daily habit for more than half of the American population.

What Claude Hopkins knew, and what every safety leader should understand, are the dynamics of habit, something now termed the habit loop. "All habits—no matter how large or small—have three components, according to neurological studies," writes Charles Duhigg in the insightful book called The Power of Habit, Why we do what we do in Life and Business. "There's a cue—a trigger for a particular behavior; a routine, which is the behavior itself; and a reward, which is how your brain decides whether to remember a habit for the future."

Hopkins, suggesting you can discover a film on your teeth by simply moving ones tongue over your teeth might be one of the most powerful cues in advertising. In fact, chances are that you have moved your tongue over your teeth at least once in reading this article. The discovery of the film would motivate action, or a routine (brushing teeth). Finally, Pepsodent was the first toothpaste to include a chemical that slightly irritated gums. This irritation is often translated into the 'fresh' and tingling feeling you feel immediately after brushing your teeth—or in the habit loop, that is called the reward.

So, why are we talking about toothpaste and utility worker safety? The dynamics that triggered Pepsodent use and the habit loop are the same dynamics repeated across the utility industry each day, for each task. But, and it is an important 'but,' we as an industry have not recognized the habit loop and intentionally tried to set this loop for our employees—it has grown by chance.

For example, the 'cue' is typically a work order, job assignment or a job popping up on a computer monitor. The 'routine' is how the work is completed and that is typically based on traditions and practices dating back decades. Finally when the work is completed our workers have a reward. The opportunity that we have as leaders is that we can intervene in one or more parts of this habit loop—we can initiate small but meaningful steps that can change habits!

Chances are we have never thought about how the cue plays into our worker's safety. What if, we substituted today's cue (the receipt of a job order) with a cue tying the work to the worker's family, for example? Or, we can focus on the routine, which is the process of actually doing the work. Here, we can take a 'cue' from Superbowl champion Tony Dungy. Duhigg writes, "So rather than creating new habits, Dungy was going to *change* players' old ones. And the secret to changing old habits was using what was already insider players' heads. Habits are a three-step loop—the cue, the routine, and the reward—but Dungy only wanted to attack the middle step, the routine. He knew from experience that it was easier to convince someone to adopt a new behavior (routine) if there was something familiar at the beginning and the end."

I think you get the point. Habits can be broken down into a discernible three-step process. For our workers, we have largely let that process develop by chance. By making a small change in the habit loop process, we can unlock great things in safety results—or we can continue to let old habits drive results—what is your reward?

Source:
Duhigg, Charles, *The Power of Habit; Why We Do What We Do in Life and Business*, Random House Publishing, 2012.

11 How to Create Your Safety Vision

A good friend called a couple of weeks ago, and he was both excited and nervous about interviewing for a utility safety manager position. If successful, he would be leading a team with more than a half dozen direct reports with influence over distribution and transmission contractors, lineman training and power houses across multiple states.

My friend wanted to write a comprehensive safety vision that would pinpoint what lineman's safety programs will look like in the next decade. We talked for some time, and this is the vision we put together.

Leadership is you. Over the last decade, the leadership focus was on senior management. Yet, an equally important group of safety leaders has been working behind the curtains to affect positive change at utility companies nationwide. These informal safety leaders often have no official title or rank, but they can make a significant difference in reducing the number of near misses and accidents out in the field.

What we will realize in the next decade is that the best safety leadership for linemen comes from fellow linemen. This influential group of informal safety leaders quietly yet effectively influence the choices and behaviors of their peers. Success in the next decade will depend on linemen teaching and coaching other linemen so that no one gets hurt.

Safety Is the New Economy—Safety is a unique mix of managing people, attitudes, training, behavior, tools, systems, policies and procedures. It's not a performance measure for the next year, but rather the performance measure for the next decade.

Utilities that can manage safety will not only be ahead of their peers in OSHA-reportable data, but will also experience a competitive advantage in several areas, such as human performance, workers' compensation, medical costs, employee engagement and worker turnover.

The economy shifted late in the last decade, and because of that, the competitive advantage through consistent safety results is more important now than ever. In the next decade, managers will stand with linemen to ensure that safety is not just a priority because priorities change; rather, it will be a value that will not shift.

Think with the Heart Rather Than Just with the Head—Utilities often fall into the same trap when it comes to safety programs. Companies often establish new rules when they discover a deficiency, and along with that rule comes a training session, a safety meeting, a set of posters, and a lot of buzz. After six months, however, the results are the same. Because there was no improvement in results, a more stringent rule is established, and the circle completes again and again.

The problem is that rules, strategies, policies, structures, procedures, monetary awards and discipline all deal in the head, rather than the heart. Policies, rules, discipline and accountability must be in place, but that isn't necessarily a ticket to results. Instead, it's just the foundation. Moving forward, utilities can achieve lasting success and sustainable results when they begin dealing with beliefs, habits, energy, passion, personal commitment and personal goals rather than just with the by-the-book regulations. For example, they can get linemen involved in the safety culture of their company through hands-on safety meetings, trusted coaching, feedback and interactive and real-time near miss programs.

The New Team

In my opinion, there is now a new definition of win-win. In the past, many utilities took an "us" versus "them" approach. It didn't matter if it was safety versus operations, management versus labor, or senior leaders against middle management—there was always tension. Moving forward, there is no scenario where one side will win and another lose; instead, we are all in this together.

In this true win-win environment, teams will take on new meaning and new energy. Organizations who understand this relationship and promote this type of environment will rewrite the definition of a team.

In the end, my friend earned this new position and his first day on the job was the first day of the new decade. As we move forward, will we act on our new vision? The opportunity is now.

12 Leadership

Thinking about What Others Think

"You can pretend to care," an old quote from an anonymous author reads, "But you can't pretend to be there." Or another management quote on caring, this one from leadership expert John C. Maxwell says, "Your people don't care what you know until they know you care."

Joanne Jaffe is a bureau chief for the New York Police Department (NYPD), and is responsible for one of the more challenging areas in the entire department. Jaffe oversees hundreds of public housing developments. While only about five percent of the city's population lives in public housing, about 1-in-5 of all violent crimes takes place in them.

As 2007 approached, Jaffe was looking, as she always does, for ways to connect with residents and to lower crime rates. She noticed a spike in robberies committed by teenagers. These crimes were mostly in two public housing areas of the city. Jaffe knew that for every juvenile arrested for a robbery, the same juvenile had probably committed an astonishing 50 to 100 crimes in which he was not arrested. In other words, if she could find a program to intervene with these youth, she could greatly reduce crimes. Juvenile Robbery Intervention Program, or JRIP was born.

Initially, JRIP aimed to mentor and monitor just over 100 of the highest risk youth who had been arrested for robbery. In December 2013, Public Broadcast Station (PBS) interviewed Jaffe. About the beginnings of JRIP she said, "The idea was now we were going to go to every juvenile (in the program) and we were going to give them a message. If they continued to engage in criminal conduct, that we were going to do everything in our power to make sure that they stay in jail. The second component to the program was really to get involved with them and their families, and identify resources to assist the family as a whole."

The program started slowly. There was no trust between those police officers assigned to the program, the youth in the program, and their families. For all of 2007 Jaffe and the officers assigned struggled to gain momentum. Then, something unexpected happened. Jaffe explains the breakthrough in Malcolm Gladwell's book *David and Goliath*.

> "There's this one kid," Jaffe said. She made up a name for him: Johnnie Jones. "He was a bad kid. He was fourteen, fifteen then. He lived with a 17- or 18-year-old sister. His mother lived in Queens. Even the mother hated us. There was no one for us to reach out to. So now, November of the first year, 2007, Dave Glassberg comes to my office, Wednesday before Thanksgiving.
>
> He says, 'All the guys, all the people on the team, chipped in and we bought Johnnie Jones and his family Thanksgiving dinner tonight.'
>
> And I said, 'You're kidding.' This was a bad kid.
>
> And he goes, 'You know why we did it? This is a kid that we're gonna lose, but there are seven other kids in that family. We had to do something for them.'"

Jaffe asked the Police Commissioner for $2,000 to purchase turkeys for the remainder of the families—he agreed. Jaffe continues the story:

> "We'd knock," she continued. "Momma or Grandma would open the door and say, 'Johnnie, the police are here'—just like that. I'd say, 'Hi, Mrs. Smith, I'm Chief Jaffe. We have something for you for Thanksgiving. We just want to wish you a happy Thanksgiving.' And they'd be, 'What is this?' And they'd say, 'Come in, come in,' and they would drag you in, and the apartments were so hot, I mean, and then, 'Johnnie, come here, the police are here!' And there's all these people running around, hugging and crying. Every family—I did five—there was hugging and crying. And I always said the same thing: 'I know sometimes you can hate the police. I understand all that. But I just want you to

know, as much as it seems that we're harassing you by knocking on your door, we really do care, and we really do want you to have a happy Thanksgiving.'"

Turkeys are a nice touch for sure, but one also needs to ask if the program worked. It did. In Brownsville, the first area where JRIP was rolled out, the number of robberies fell 77 percent between 2007 and 2011. The JRIP officers also took their first Thanksgiving experience and let it drive other interactions such as toy drives, summer basketball leagues and activities among families who are in the JRIP program. In fact, two years later the NYPD Commissioner expanded JRIP to other parts of New York City and other major cities are looking to implement this type of program.

So, what does this have to do with employee safety? Results in employee safety have much to do with relationships; and leaders can find a number of take a ways from this story. Here are five key points to remember.

Grassroots—In a 2009 press statement, NYPD Commissioner Ray Kelly said, "This type of grassroots policing is essential to reducing and ultimately preventing juvenile crime. We have already seen success from JRIP and we expect that will have similar results working with young people and their families in East Harlem." The heart of this program is building a grassroots program, which in effect extends the police department. Are you extending your safety department?

Breakthrough—Jaffe has a strong emotional tie to what could be called the program's breakthrough, the 2007 Thanksgiving Turkey program. It actually started when front line officers took a proactive step to buy a turkey for the highest risk of the high-risk youth. And the overall 2007 Thanksgiving Turkey program was driven, for the most part, by front-line officers with strong support from executive leadership. Where will your breakthrough come from?

Compassionate Accountability—JRIP works because of the strong accountability for those youth enrolled. If you screw up, you go to jail. But, that side of the program is balanced with the compassion of the officers who work it. They take youth on job interviews, check on grades, and otherwise provide opportunities and mentoring that they would otherwise not have. The program works because there is both compassion and accountability. Is that true of your safety accountability?

Innovation—If JRIP had been put to a public vote before it was allowed to be executed, I'm guessing it would have failed miserably. Think about it, enroll the toughest juvenile offenders into a close watch program and provide mentoring and services that other youth, who have not been arrested, don't receive—and oh yah, a Thanksgiving turkey too! It took a lot of courage for Jaffe and other leaders to write, then execute, a plan they knew was unorthodox, but one that would also yield results. "You can't solve a problem with the same thinking that created it," reads an old saying. Are you innovating your safety program?

Caring—we will close where we began. "You can pretend to care," an old quote from an anonymous author reads, "But you can't pretend to be there." Or another management quote on caring, this one from leadership expert John C. Maxwell says, "Your people don't care what you know unit they know you care." Do your people know you care?

Today, spend some time thinking of what others may think.

Source:
Gladwell, Malcolm, *David and Goliath: Underdogs, Misfits, and the Art of Battling Giants*, Little, Brown and Company, New York, 2013.

13 How Leaders Think about Near Misses

Turning Remote Misses Into Near Misses

Traffic was heavy, cars were dashing by. I was in the left hand turn lane, my turn signal flashing. The traffic light was green, but it was a left turn yielding to oncoming traffic. In my nearly three decades of driving I had negotiated thousands of these turns. I saw my chance. There was a car barreling down, but enough space for me to speed through. I moved my foot to the accelerator but I flashed back to a near miss a month earlier and I stopped. The car barreling toward me sped by and I just sat, thinking and waiting for a safer option.

It was time for another safety meeting. It was 2003 and I was a first line supervisor of electric distribution crews. Typically I would prepare and lead safety meetings. As I laid out material for the meeting an email popped up. It was an update on a fellow line worker in another division. Unfortunately as this worker was working one night during storm recovery he was struck by a car. The injuries were severe, life changing, and included massive head trauma. The email was an update on his condition.

I recall vividly that I closed the safety meeting that day with the email; an update on our coworker's condition. All of us had been keeping him and his family in our thoughts. We had also 'passed the hat' and collected a significant amount of money for the family to help offset expenses during this tough time. As I finished the safety meeting, I urged everyone to follow all work zone traffic rules including advanced warning signage, proper lane closure if needed, flagmen if needed, and use of high visibility vests.

Later that day I was out looking at jobs crews would be doing the following week. This was rural Missouri and the tree lined roads were a mix of hills and curves making visibility poor despite the 55 mph speed limit. As I drove around a curve I found one of my line workers in the middle of the road holding his hand out stopping traffic. His crewmate was working some equipment along the road that was taking one lane of traffic.

I pulled over and got them and their equipment off the road. I couldn't believe this. Just five hours earlier we shared an update on a coworker who was hit by a car under very similar circumstances. The result of that incident was life changing, tragic. But here there was no advanced signage, no established work zone, cone tapper, flagman, or use of high visibility vest. I demanded to know what they were thinking. The reply was simple, unemotional and confident, "No one is going to hit us out here."

April 24, 2014 was a Thursday; and it started better than most days. I was able to have breakfast with my family and was running early, which gave me time to roll through McDonalds for a $1 iced tea before settling in for the normal 30 minute commute, which is the stretch of Highway 63 between Columbia and Jefferson City, Missouri.

I was on cruise control with my mind thinking about the day ahead. Out of the corner of my eye I saw a white car entering the highway at a fast pace. On this particular stretch of highway there wasn't an on or off ramp; it's a grade-level crossing where a driver crosses over the northbound lanes, yields in the middle, and then proceeds across the southbound traffic. The driver of this car raced across the northbound lanes, barely yielded, then dashed into the southbound lanes. I quickly reasoned that she eyed the two cars in the southbound passing lanes and was intent on beating them across the highway. She did. But, in focusing on getting ahead of them, she didn't see a black Ford Expedition in the southbound driving lane. The impact still burns in my mind.

The crash occurred right in front of me. I pulled over. I called emergency services. I got out of my car. Steam and smoke misted from what was left of the engine compartment as the radio in the crashed car still played. I was joined by a handful of other people, a few wearing scrubs—that was a clear sign they knew more about medical aid than me. The woman in the car was lying lifeless across the front seats.

One person checked for a pulse and for breathing. Given she had both we decided not to move her. We waited. I prayed.

Later that day a friend sent this blurb to me; "One woman was killed and another seriously injured in a Thursday morning crash on Highway 63 near Ashland. A 19-year-old driver was crossing Highway 63 in a Dodge Avenger and failed to yield to 34-year-old who was driving a Ford Expedition. The SUV hit the car on the passenger side. The 19-year-old driver was pronounced dead at University Hospital."

So why were my flash backs to the car crash causing me to change my behavior and drive more cautiously, while linemen who had been praying for and financially supporting a coworker who had been hit by a car not change their actions at all? It has to do with the fact I was impacted by a near miss, the linemen were impacted by a remote miss. Let me explain.

As World War II was beginning, British leadership was very concerned about the German's bombing of London. Why? Well, there is the obvious loss of life and damage to the city. But England's leaders thought the effects of the bombing would literally shut down the city. They were certain that bombing night after night for weeks or months on end would cause all Londoners to stay in their homes and bomb shelters. The British government needed workers in factories and in businesses so they could continue to manufacture what was needed to fuel the war effort and keep the economy going. And, Germany did bomb London night after night for two and a half months. And, what happened? Actually, the opposite happened.

A number of researchers and scientists have studied this reality trying to explain it. One of the best and most widely accepted and influential theories is from Canadian Psychiatrist J. T. MacCurdy. MacCurdy asserts that when an event happens, bombing in this case, the population can be divided into three groups. The first group in the case of the bombing are those directly hit. Unfortunately, they were killed or severely injured thus hospitalized. Due to the result for this group, they were not running through the streets telling everyone how bad the bombing was and to take cover. Due to their injuries they were unable to do that. As MacCurdy says, "The morale of the community depends on the reaction of the survivors." The next group MacCurdy called the near miss group. A near miss leaves one 'deeply impacted.' In the case of the London bombings, these would have been people directly involved with a bombing but not hurt. Maybe they helped pull injured from ruble or pulled bodies from a building. Maybe they witnessed a serious car crash and stopped, or administered first aid to their coworker after he was struck by a car, before the ambulance arrived. The final group MacCurdy called remote misses. Those in this category may have heard the bomb sirens and even heard a bomb hit, but it was in the distance and they were not directly effected. In these cases the affect is the opposite of the near miss group. A remote miss leaves one feeling invincible, or as MacCurdy wrote, "Excitement with a flavor of invulnerability."

In fact, MacCurdy interviewed a number of people from the London bombings to reinforce his theory. One women, for example, said, "When the first sirens sounded I took my children to the dugout in the garden and I was quite certain we were all going to be killed. Then the all clear went without anything having happened. Ever since we came out of the dugout I felt certain that nothing would hurt us."

"Nothing is going to hurt us," sounds very similar to what the lineman told me when I found him in the road directing traffic without any of the proper work zone traffic protection or PPE.

The Near Miss report has been a long-term staple of the utility industry, but I wonder if more times than not it has been the 'Remote Miss' report. The standard format for these reports is simple, the report generally described conditions prior to the incident, the incident then transitions to recommendations to prevent reoccurrence. But, as MacCurdy explained, for a near miss to change behavior, like yield a little longer on a left hand turn for example, one must be 'deeply impacted' by the incident. Unfortunately, our workers are like Londoners. The German bombing of London resulted in 40,000 deaths and 46,000 injuries. But, spreading that across a metropolitan area with more than 8 million people means there were many more remote misses, who felt invincible afterwards, versus near misses—people who were deeply impacted and changed their behavior. Are injury near miss reports any different? Sharing a dozen near miss reports over the course of a year to hundreds, if not thousands, of workers working thousands, if not millions, of collective hours per year within an organization means we are creating remote misses, workers who feel even more emboldened to stand in street without PPE and work zone protection because, "No one is going to hit us out here."

Knowing this, we can plan differently when sharing these incident occurrences. Here are some things to consider:

The primary reason to share these reports is to change behavior (getting people in the right work zone traffic protection, for example). If we listen to MacCurdy we need to 'deeply impact' our workers in order to have a change in behavior. There are several ways to intensify 'impact.' For example, spouse and/or older children can write a note asking the employee to comply with the rule. Or, have the injured worker give his personal story, versus a supervisor reading it from a page.

Next, think about making it an involved safety meeting activity, versus a 'sit and listen' reading of an incident summary. For example, in the case of work zone traffic protection, design a refresher of the subject including giving workers paper and markers to draw the proper work zone protection based on street and road conditions scenarios you give them. You can then close the meeting with the incident report—it is part of the safety meeting reinforcing the refresher training, and not a stand alone to create a remote miss.

Incident reports needs to be shared, but knowing we risk creating hardened Remote Misses means these reports are not just a simple 'reading' but more . . . what more? Just enough to deeply impact those you are sharing with.

Source:
Gladwell, Malcolm, *David and Goliath: Underdogs, Misfits, and the Art of Battling Giants*, Little, Brown and Company, New York, 2013.

14 How Safety Leaders Win

If you were asked to pick the best, most successful, college coach who would you suggest? UCLA's legendary basketball coach John Wooden comes easily to mind, as does Alabama's football coach Bear Bryant. Thinking a little more, Mike Krzyzewski Duke's basketball coach is a good guess as is Tennessee's former woman's head basketball coach Pat Summitt. Or, Dan Gable of Iowa wrestling fame is certainly on the short list. But a name that you would probably not have said is Anson Dorrance. Coach Dorrance is the head coach of one of the most successful programs in the history of college athletics. He is the head coach of women's soccer at North Carolina.

Coach Dorrance's coaching record is almost unbelievable. In 1979 Dorrance worked with the Association of Athletics for Women to establish a national women's soccer program. The first NCAA women's soccer championship was three years later, won by Dorrance's North Carolina Tar Heels. For his 33-year career as head coach of the Lady Tar Heels, Dorrance is 719–39–24 for an unheard of 93.5 percent winning percentage. He led his teams to a 101 game winning streak and coached 20 players recognized as National Player of the Year. If that's not impressive enough Dorrance's teams have won 21 NCAA championships.

So, what's Coach Dorrance's secret? His success over time is likely much more complicated than can be shared in a short 1,200-word article. Having said that, there are three keys to his success that can provide great insight into his winning program, and also give us additional insights into how we can be even more effective in leading safety programs.

Compete

There is a legendary story about Henry Ford. He had two shifts at one of his factories. These two shifts used the exact same assembly line so had the same tools, work hours, number of employees, etc. For some reason the first shift always assembled more cars than the second shift. One evening Ford walked on the floor of the second shift, took a piece of chalk and wrote a number on the floor, 48 for example. The number represented the number of cars produced by the first shift. Ford didn't say any words, he just wrote the number and left. When he walked in the next morning he was surprised to see his number was scratched out and a new number written—54, that was the number produced by the second shift. Ford left it there and the first shift then wrote next to it as their shift ended, 56. The night shift would not be outdone—they produced 58 that next night.

Coach Dorrance says, "Competition is key to developing players. The only practice environment in which you truly develop a player is a competitive arena." What most teams consider drills, Dorrance will chart then post for all of the players to see. It pushes those players who are on top of the list to stay there, and those who are not on the top of the lists to work even harder.

I do not support the idea of competing for the lowest injury numbers since this could drive workers not to report incidents, which is not the goal. But, this idea of competition is one to consider for training exercises, safety rule knowledge, most safe acts, most supervisory job observations, housekeeping inspections, and many more categories. Post these lists and reward those on top, and your organization will push to be even better.

Care

A December 7, 1998 *Sports Illustrated* article says that, "Dorrance insists that players call him Anson. Before the last home game of the season, he presents each senior with a red rose. On the wall in his office is this sign: *PEOPLE DON'T CARE HOW MUCH YOU KNOW UNTIL THEY KNOW HOW MUCH YOU CARE.* "That is the critical element in coaching women," Dorrance says. "*You can't get me to do anything unless you care about me first.* Soccer is not that important to them. Connection is."

When I was a safety professional serving a major Midwestern utility I served over 400 electrical line workers and substation technicians. One thing I did to make a connection is to write a personal birthday note to each worker and supervisor, about 500 letters per year. After a few years of this I saw a birthday note that I had written taped to a locker. I found the locker's owner and asked him why it was there. He simply said that in nearly 30 years working for the company receiving that birthday note was the best thing that had happened to him.

Dorrance says, "This game rewards you for playing with huge hearts." For those workers who had received birthday notes, they simply knew that I cared. So when I showed up to coach, teach, instruct or correct, I mostly found open ears and workers willing to listen—because they knew I cared.

"He taps into the core of your being," says national team assistant coach and former Carolina player Lauren Gregg. Anson Dorrance cares.

Prepare

Dorrance wants soccer games to be a 'day off' for his team. What he means by that is he wants practice to be so hard and 'tough' that his team is very well prepared for any game. Practices are so stressful that players have named Tuesday's conditioning day as 'throw-up Tuesday.'

"We've tried to design a system that's difficult to play against," Dorrance said. "That system is predicated on work ethic and high pressure. It's hard for other teams to replicate that in practice. Often times, even when a quality team plays us for the first time, it's a bit of a shock."

How do our work groups prepare? And as supervisors and managers, how do we prepare them? Are our training sessions, job briefings and tailgate sessions, safety meetings, safety committee work really preparing our workers for the day's hazards?

Dorrance says, "The vision of a champion is someone who is bent over, drenched in sweat, at the point of exhaustion when no one else is watching." Does your safety program do the equivalent?

In the end, if we can infuse our safety programs with healthy and appropriate competition, caring and preparation, we will find our results to be elite—just ask Coach Anson Dorrance.

15 Pashtunwali: A Safety Code

Are We Really Our Brother's Keeper?

All Marcus Luttrell could think about was the thirst. Marcus, a Navy SEAL, had been shot at least two times, had shell fragments in his legs and had broken vertebra from a fall. He had lost his pants in the most intense and deadly day for U.S. Special Forces since World War II. But, 36 hours after it all began all Marcus could think about was the intense thirst.

Marcus Luttrell was part of a four man SEAL team dropped in remote Afghanistan on June 28, 2005. These four SEALs were part of a larger mission called Operation Red Wings. The operation targeted an elusive militia leader affiliated with the Taliban named Ahmad Shah. Shah's group was causing all kinds of havoc for U.S. troops in the region and Luttrell's team had orders to find him, and stop him.

Luttrell, along with his fellow SEALs Petty Officer Danny Dietz, Petty Officer Matt Axles and Lieutenant Mike Murphy were dropped in the cover of darkness in the early hours of June 28th. After hiking for several hours through treacherous terrain, the four dug in on a ridge as the sun was getting ready to break the horizon. They waited. Still early in the morning, a shepherd jumped from a log nearly landing on Luttrell's gun, the shepherd had not seen Luttrell. There were two more herders and about 70 goats—the mission was compromised. The SEALs didn't have many options but to let the shepherds go. About an hour later Taliban militia forces ambushed the SEALs.

The four Navy SEALs, greatly outnumbered, fought. They tried to radio for back-up but couldn't get a signal on the satellite phone. The SEALs were pushed down the steep hill. Danny Dietz was shot and Luttrell would drag him down

Pashtunwali: A Safety Code

the hill, set him up to fire, then move again. As Luttrell went to move him again Danny was shot in the head and died instantly in Luttrell's arms. Later during the firefight Matt crawled over to Luttrell and said, "I'm sorry, bro, I can't help you because I'm blind. They shot me in the face. Luttrell would later be separated from Matt as an RPG landed close to them as they were huddled together, Matt was thrown one direction, Luttrell another. Luttrell never saw him again.

Lieutenant Mike Murphy knew he had to take action to get help. He calmly walked into an open clearing where the Sat phone could receive a signal, but where he would also be an easy target for the Taliban fighters. He called for help. Told those who received the call 'thank you' and was shot and killed.

As it grew dark Luttrell was all alone. He had been shot at least twice. Had fragments in his legs and broke his back pin balling down the steep embankment. He found a hole, packed his wounds with dirt to slow the bleeding. The Taliban was hunting him.

It was thirst that drove Luttrell to move, to crawl in search of water—any moisture. Luttrell didn't know that U.S. Special Forces were searching for him and his fellow SEALs. He didn't now that a rescue mission had been scrambled almost immediately after Mike's call. Luttrell also didn't know that two Chinook helicopters with Special Operations Forces on board raced to the mountainside where the four SEALs had been fighting, when one of the Chinooks, with eight SEALs and eight Army Night Stalkers was shot down killing all on board.

Some 36 hours after the first bullets were fired, Luttrell crawled to a spring, he had found water. He put his face under the stream when he caught site of men—Luttrell was surrounded. Given his SEAL training, he grabbed his gun and his last grenade and threatened to fight. The men told Luttrell they were not Taliban, and would help him. Exhausted from the hours of fighting, and badly wounded, Luttrell put down his weapons.

The men, led by Mohammad Gulab took Luttrell and provided care. They hid him, moving him frequently so the Taliban could not find him. Taliban leaders scoured the village for Luttrell but could not find him. And, they threatened Gulab, saying they would kill his family if he didn't turn over Luttrell!

Gulab was later asked why he did what he did, risking his life for a total stranger. Gulab simply said it was *Pashtunwali*. Pashtunwali is a tribal custom of respect. "A respect," Gulab says, "for a guest that comes knocking at your door. And even if he is in need, or if he is imminent danger, we must protect him. I knew I had to help him, to do the right thing, because he was in a lot of danger."

In the utility industry there is a longstanding tradition of being our 'brother's or sister's keeper.' What it means, much like Pashtunwali, is that we will protect those entrusted into our care. But, unlike Gulab and his code of Pashtunwali, all too often we fail. Today, it is not uncommon that we look the other way if a seasoned journeyman is breaking a rule or taking a shortcut. Today, we often allow a practice that is 'close' to how it is supposed to be done, but not exactly how it is supposed to be done. Today, we understand that there is what we learn training, and the way we do things in the field . . . and this list can go on and on.

When a utility brother cuts a corner, he is in imminent danger, and we must protect him. Not by allowing the at-risk act to continue, but by paying that worker the respect he deserves. We must stop the job and get it right. Pashtunwali would call for no less than this.

Luttrell entrusted his life to Gulab, just as we entrust our lives to our coworkers. After four day of hiding Luttrell from the Taliban, U.S. forces rescued him from Gulab's care. Without Gulab, and without the tribal code of Pashtunwali, Luttrell would have most certainly died.

For his bravery Marcus Luttrell was awarded the Navy Cross in a White House ceremony. Matt Axelson and Danny Dietz were also awarded the Navy Cross posthumously. For sacrificing his life to make that telephone call, Lieutenant Mike Murphy was given the Medal of Honor. His parents accepted it. It was the first time the country's highest military honor was awarded for service in Afghanistan.

Pashtunwali: A Safety Code

Navy Commander Joe Maguire who was the top ranking officer in charge of SEAL training when this occurred said, "These are just you know unremarkable men who do absolutely remarkable things." May others say the same about us when it comes to keeping our brother's and sisters's safe.

16 How Leaders Move from Soft Skills to SMART Skills

In safety, there is an old saying that says, "The hard stuff is easy but the soft stuff is hard."

The 'hard' stuff for safety professionals, managers and supervisors are the things we can see, read and touch and because we can see, read and touch them they are easy. These 'hard' things include safety rules and safety manuals, OSHA compliance documents and directives, PPE, monitoring, etc. In short, if an employee comes to ask you a question and you can answer it from a book, company policy, picture, or other resources, it falls on the hard side of safety . . . and this is the easy side.

We all know the expression, 'One can lead a horse to water but you can't make him drink.' Just because our employees know what is written in a safety manual or expected by an OSHA policy does not mean that those rules are followed. To gain maximum compliance with the rules and policies (hard stuff) and to build a sustainable safety culture safety professionals, superiors and managers must also be well skilled in what some call the 'soft' skills of safety. These are the skills that can lead people to the safety rules and gain compliance. But, I don't think 'soft' adequately describes them. Instead, they are SMART skills, and leaders—and the top achieving organizations—know and understand SMART skills well.

To begin SMART skills can be defined as the intangible human leadership behaviors geared toward positive engagement of employees, with the goal of trust in order to gain long-term understanding, thus rule compliance. How do we get there, to these SMART skills? Consider the following.

S: Spirit—*Energy Drives Results!*

There is an old story that goes like this, "There was a man walking through town and he saw a buddy over on the bridge getting ready to jump. The man ran over and encouraged his friend to come down and talk. He did come down off the bridge and they did talk . . . and two hours later they both jumped!"

The point is simple, no one likes to be around an energy drain. As a matter of fact, it's just the opposite; we need to bring energy into each day, each situation, each interaction. Jim Loehr and Tony Schwartz in their book titled, *The Power of Full Engagement; Managing Energy, Not Time, is the key to High Performance and Person Renewal* shared the following three concepts in their book. First, "Energy, not time, is the fundamental currency of high performance. Performance, health and happiness are grounded in the skillful management of energy." And finally, "Leaders are the stewards of organizational energy!" Do we save our energy for after our work shift, or do we bring it everywhere we go, including work? Energy gets results—SMART results.

M: Magnetic—*The Ability to Draw Your Team to You in a Positive and Powerful Way*

Take a moment to work through this simple five question quiz. First, for each of your direct reports, how many have children and what are their children's names? For your direct reports, who is married and what is their spouse's name? Do you know the primary interest or hobby of each of your direct reports? In the last 30 days have you given direct feedback about work performance and/or safety to each of your direct reports? Finally, in the last 90 days, have you done something 'special' for each of your direct reports, like a note on their birthday, a thank you card for a job well done, etc.?

Randy was a tough, hard to get to know, son of a gun. He had more than twenty-five years of experience and didn't get along the best with the local management. I knew Randy and had actually worked with him when I was an apprentice. I was now the safety supervisor for the area. In this role I was responsible for nearly 400 linemen, substation workers and gas employees in rural Missouri. Knowing that I couldn't see everyone in the course of a month, I decided to show each I cared in a different way. I got with Human Resources and received a list of everyone's birthday. Then, I wrote each person a birthday note. For Randy I said something like, "Happy Birthday, I really liked working with you back in the day. I always liked your funny stories. Work safe, Matt." I sent the note and forgot about it.

Family counselor and author Josh McDowell wrote, "Rules without relationship causes rebellion." About five years later I happened to be in Randy's show-up location and there on his locker with the pictures of his wife and kids was a faded piece of paper. I recognized it immediately; it was my note to him. I found Randy and asked why the note was on the locker with family pictures. He smiled and said, "In all of my time with the company, that was the nicest thing anyone has done for me."

If I needed to talk safety or enforce a rule with Randy, I'm confident he would have listened because he knew that I cared—I had drawn him in. Changing behavior is first about creating a relationship . . . being magnetic.

A: Accountable—*Answering the Question "Who do you work for?" Each Day*

A quick quiz, who do you work for? For most, the automatic response is the name on the paycheck. Others will say, 'my boss.' After some thought, a few recited 'family.' The truth is we work for ourselves. We trade time and talent to an employer for money. We are each CEOs of our own business. We each have a corporate budget, the money we have to spend, a corporate fleet, the vehicle we own and drive, and a corporate board, our family and friends. This fact is important because if I'm hurt at work I suffer, not my employer. Sure the employer will pay a financial cost for that injury, but the great secret is that the employer will continue to make money. The injured cannot make another eye, hand or finger. Your daughter cannot make another dad; your dad can't make another son.

In April 2006, the Centers for Disease Control (CDC) released a report that listed the lifetime costs of the workplace injuries that occurred in 2000; the costs topped the $406 billion mark. Those numbers include the employer paid medical expenses ($80.2 billion), but the bigger numbers ($326 billion) came from lifetime productive losses that include, "lost wages, fringe benefits and ability to perform normal household responsibilities." (CDC 2006) Those losses aren't felt by the employer but by each of our people when injured.

Recently, the National Safety Council awarded UPS Chairman and CEO Michael L. Eskew with the coveted Green Cross for Safety Medal. Eskew says this about safety, "But for safety to be a core value, it has to be taken personally." (McMillan 2007) Instilling in each of our people exactly 'who they are working for' is the first step in personal accountability.

R: Risk—*Doing Something That Is Positive and Powerful, for You and/or for Another That You are a Little Nervous to Do*

You can pretend to care," reads one of my favorite quotes from an anonymous author, "but you can't pretend to be there." Bruce was a minister and researcher. He approached a charitable board foundation a number of years ago. This foundation had a reputation for awarding grants for research on human effectiveness. Bruce's proposal was that he would tour the United States and Europe and interview the most successful business people and politicians of the time; asking them what was the one key to their success. The foundation awarded Bruce a grant and for two years Bruce spent hundreds of hours talking to these successful people. He drilled down on the one key to success. When the two years was over, he returned to the Foundation to report his results. And, what did Bruce tell the Foundation was the one key to success? Risk.

Now, I'm a safety professional by trade and training and I'm not advocating a 'risk' that can get someone hurt; nor was Bruce. Bruce understood that people could misconstrue his study results and take unwise or careless risks all in the name of working toward success so Bruce put risk into categories. One category was emotional risk. I have written a definition for emotional risk, *when you do something you are a little nervous to do, that is positive and powerful, for yourself or someone else.*

Each day safety offers a unique opportunity to take emotional risks to coach employees, to redirect at-risk acts and to reinforce strong habits.

T: Trackable—*How do You Keep Score?*

"Your people play harder when they know the score." Just because these skills are 'SMART' and are more difficult to put our fingers on compared to the 'hard' parts of safety does not mean that we shouldn't set up a number of comprehensive tracking and measuring processes so that you, your team and your senior executives 'know the score.' SMART skills include the human intangibles and can be broken down, tracked and measured. Consider a monthly dashboard that supervisors, managers and your senior leaders can access. In that report measure the traditional 'hard' items like injuries, yet in the SMART section look at job observation reports, critical conversations, crew communication, and even birthday notes delivered, if you like.

17 You Can't Pick the Time

One of my favorite past times, and a favorite of countless millions of other people too, is filling out the annual NCAA men's basketball bracket. And, if one really spends some time analyzing each team, coach, player and match up, you can actually come pretty close to picking a perfect bracket, right?

In March of 2014 billionaire Warren Buffett and Quicken Loan founder Dan Gilbert added a little 'extra' incentive to the annual bracketology exercise... and by 'little extra' I mean a $1 billion dollar prize for anyone who submitted a perfect bracket.

The offer was simple. Anyone who was able to pick all the winners in all of the games in each of the rounds would be rewarded with his/her choice of $25 million over 40 years or a lump of a cool $500 million—winner's choice.

Thinking about the path to a perfect bracket starts with the first round of the tournament, which has 64 games. Each game would have one winner. Each team has a ranking, for example the first seeded team plays the 16th seed. Easy picking right? The next round has 32 games, then 16, etc. On paper, it seems fairly easy but in reality the odds are stacked against you—why do you think Mr. Buffet is willing to put up a billion dollars? When one adds up the odds of picking each game correctly for each round the numbers stack up against you and your picks—big time. The odds of picking a perfect bracket is a mind-boggling one-in-9.2 quintillion. If one was to write out that number it would be nine with eighteen zeros.

Unfortunately, too many of our workers sometimes take a small chance or a quick shortcut reasoning that they will not get hurt—it's the same as picking the 5 seed over the 12, the 5 always wins, right? Actually, the 12 seed wins one of three games over the much higher 5 seed. Which, by the way, is the same odds as an 11 seed beating the higher 6 seed. Coworkers and supervisors sometimes see at-risk acts but make excuses that, "he knows what he's doing," or "he's a journeyman so he won't get hurt." In reality reasoning you won't get hurt is like thinking that you CAN pick the entire NCAA basketball bracket—the odds are stacked against you. The odds are against you or anyone taking a shortcut on not getting hurt.

This year it only took the first 25 games to eliminate all of the thousand and thousands of participants who submitted brackets in the billion dollar bracket challenge. And, a 7 seed, University of Connecticut, won the entire tournament. I can't tell you when you or a coworker will get hurt if a shortcut is taken, but I can tell you that you will get hurt. Those odds are just as certain as not picking a perfect NCAA bracket. I'd bet a billion dollars on it.

18 Always Take the Shark Training

Five Things Linemen Should always Do

"Prepare to crash," were the only three words the pilot, Phil Phillips, uttered into the intercom on The Green Hornet, a United States B-24 staffed with eleven crew members in August 1942. Moments before these words sounded one of the plane's two right engines had stopped. In an effort to 'feather' the dead engine, the plane's engineer accidently killed the other left engine. Now both engines on the left were dead. The crew was only 800 feet above the Pacific Ocean performing a search and rescue operation. Phil had only seconds to restart the one good left engine in order to balance the plane. He worked frantically to do that, but in moments he realized it was no use. He made the call.

Louis Zamperini was the crew's bombardier. A second lieutenant, he left University of Southern California where he was on an athletic scholarship to join the Air Force. Just months earlier Louis and Phil had flown the lead plane in a 23-plane squadron in a bombing raid on a Pacific Island named Nauru. There, they took heavy fire from the Japanese but were able to drop their bombs. The return was tricky. They had countless holes in the aircraft and no brakes to stop the plane. They performed an emergency landing and miraculously skidded in safely. Upon landing they counted 594 bullet holes. Now, Louis used these precious seconds to make sure everyone was ready for the crash. He handed out life vests and grabbed one for himself. He manned his assigned crash position. His mind throbbed with one single thought; *nobody's going to live through this.*

850 miles west of Oahu The Green Hornet shattered in the cold sea. Louis was trapped by wires that dragged under water. He passed out and immediately regained consciousness only to be free of the restrictions. He pulled the cord on his

life vest. On the surface he found Phil and Francis "Mac" McNamara in a life boat. He joined them. No one else survived.

On the life boat, there, with little water and even less food, they floated. Days passed and then weeks. In addition to the lack of food and water, there was real danger from sharks. Sharks of all sizes circled the life raft day and night. In fact, several times while lost at sea, Louis found himself in the water and face to face with a large shark. The shark came after him, but Louis remained calm and knew what to do. He had been trained to repel sharks.

Before the war, Louis Zamperini was a track star. In 1934, as a high school senior, Zamperini set a world interscholastic record for the mile, clocking in at 4:21.2 and a week later he won the California state championships. He enjoyed college, but his real goal was the 1936 Olympics. He earned a place on that team, qualifying for the 5,000 meter event. At 19, he still holds the record for the youngest qualifier for that event in US history. While he didn't win a medal, he finished eighth. The mile and 5,000 meter are dominated by older athletes, those in their mid- to late-twenties. Given what he did at age 19, his Olympic future was bright.

As an airman, one might think that Louis would have traded air skills for Olympic training, he did not. Louis threw himself into this role as both a bombardier and a leader on the crew. To that end, when an Island Native offered a shark training class, giving tips to repel sharks if lost at sea, Louis signed up. But the interesting thing is that of the nearly 30,000 airmen who could have signed up, only 28 actually took the class. Months later, floating in shark infested waters these skills were used by Zamperini to keep him and his fellow survivors alive as they were face to face with a hungry shark more than once.

In addition to the deadly sharks, the men struggled with severe hunger and thirst. McNamara died after 33 days at sea. On the 47th day, Louis and Phil landed on the Marshall Islands. It is believed that these two hold the record for the most days survived 'lost at sea.' Both Louis and Pete, who were captured by the Japanese when they landed on the Marshall Islands, survived their brutal prisoner of war experience and made it home alive. Making it home alive took skill, luck and faith. It also took training, and in the case of Louis Zamperini, shark training. Just as I would advise you to always take the shark training, here are five things that as a utility worker you should always do.

Always Know the Rules—This sounds so basic but not everyone knows all of the rules. Some years ago when I was working as a safety supervisor supporting nearly 400 linemen, I pulled up on a job. There six linemen with over 100 years of combined experience were working 34.5 KV de-energized and grounded. Their rule book required supplemental grounding, which they had not installed. I stopped the work and called the crew together, not one of them knew this rule. If a doctor hurt or killed a family member because he didn't know a medical rule it would be unforgivable; after all, the doctor is a professional. Utility work is no different.

Small Stuff Matters—No doubt when you read the previous section, "Always Know the Rules," you thought of the 'big' safety rules like rubber gloves, grounding and checking for dead. And you should think of those, they are vital to a utility lineman's safety. But, small rules are important too. Just like taking the shark training, do you check your rubber gloves or rubber goods before each use? Do you chock your truck? How about setting the parking brake on your truck? Be one of the 28 out of 30,000—do the big stuff, and then take the shark training too.

Safety Stop—In the defensive driver class for utility trucks, there is specific instruction for executing a right hand turn on a red light. First, the driver shall stop completely and scan the intersection. If it is clear, he may execute a right turn. But, after moving forward a few feet, stop one more time, a safety stop. On this second stop, the intersection is scanned again, and if clear then it is safe to proceed. A safety stop isn't just for right hand turns on red. Before any safety sensitive task such as climbing a pole, gloving primary, lifting a jumper, etc., perform a safety stop. This can be as short as five seconds, to make sure that everything is safe before you proceed. Nothing bad happens if you first stop—even if it's for just a second!

Plan Safety—Electrical line work and utility work is so hazardous that OSHA requires that before any work begins that there is a job plan. Each time, before each job the crew must review job hazards, safe work rules, special precautions (hidden hazards), energy source controls and PPE. Too many times a crew will talk about the work at hand, but won't cover these five basic things. When I was an apprentice, there was an old crew foremen. He would always take the time to plan the job, and cover these key areas too. After the job briefing he would say, "and remember boys, no one gets hurt today." When we plan, and when we take the shark training, no one gets hurt today.

Ask for Help—Louis, Pete, and Mac were on their own, drifting. Often, as a utility line worker we may feel like we are on our own. I remember my first call out as a journeymen. It was a simple transformer outage. But it was dark and rainy and in an area that I was not familiar with. After what seemed like hours of troubleshooting, I still couldn't figure out the problem. I felt very alone, like it was all on me. But I was not alone. I had the dispatcher, my supervisor and all of my fellow linemen who were just a phone call away. Ask for help, you are not alone.

Source:
Hillenbrand, Laura, *Unbroken: Unbroken: A World War II Story of Survival, Resilience, and Redemption*, Random House, 2010.

19 Four Key Steps to Conducting an Effective Job Briefing

In the first five months of 2008, the tower industry was saddened and rocked by seven tragedies. The seventh occurred on May 22 as a young man named Joe Reed was climbing a tower outside of Miami, Florida. He fell off the tower and lost his life that day. In response to these workplace fatalities, the National Association of Tower Erectors took the lead by sponsoring a webcast for more than 320 of its members. The message was simple: Hazards must be identified, communicated, then controlled and managed. The best way to do that is by identifying hazards through a job briefing or tailgate session before work begins.

In addition, OSHA outlines five key elements to be covered in the briefing: hazards associated with the job, work procedures involved, special precautions, energy source controls, and personal protective equipment (PPE) requirements.

The only problem with OSHA's outline is that it can be very hard for linemen to remember in the field. Here is a new approach for job briefings: the S.A.F.E. model. As always, please compare this model to your respective safe work rules. If possible, however, use the following tips to help ensure that you can go home safely at the end of each day.

See and Say All of the Hazards—A few years ago, I sat in a safety committee meeting with a substation group as my cell phone rang. Because I wanted to pay attention to the meeting, I sent the call to voice mail. The phone immediately rang again, and I reluctantly answered. It was the regional dispatcher. "Electrical contact Matt," the voice said. "A boom contacted an overhead line. Two men are down."

As I arrived on the scene and gathered facts, I learned that a crew had positioned a truck near a line to complete a routine job of setting a pole and transferring phases. The first task was to simply frame the pole. Unfortunately, in the process, the boom contacted the line. The overhead line was the one and only hazard on the job that could have drastically changed or ended a life.

In the end, a good person lost his life on that day, and the lives of the crew, a work group, and families were changed forever. This incident could have been prevented by seeing and saying hazards before work began. Before beginning a job, linemen must ask themselves, "What are the three primary hazards?"

Ask "What Did I Miss?"—Anyone who has climbed a pole, exposed underground cable or driven a staple has both heard and told stories regarding hidden hazards. These hard-to-find hazards are everywhere on our job sites.

Linemen and utility workers must find and eliminate these hazards. After identifying any of the obvious hazards, they must next dig deeper, ask probing questions and consider such factors as approaching darkness or changing weather or traffic conditions.

Follow All Rules—It seems that the utility industry as a whole has prescribed to the profound words of Larry the Cable Guy, "Git-R-Done!" After all, we know we have a lot of work to do, so we don't feel the need to sit around and talk about it. Yet, that is exactly the point. After we have identified the hazards and probed deeper for hidden hazards, we need to take the next step and talk specifically about safe work rules, procedures and PPE. There will be plenty of time to "Git-R-Done" once everyone on the crew is on the same page.

Pay Attention to Energy Source Controls—With all of the hazards and rules that apply to line work and utility work, have you ever taken a moment to consider why OSHA set energy source controls as a stand-alone category within job briefings? They did so because of the life changing or life-ending effect of a mistake or shortcut in this category.

Four Key Steps to Conducting an Effective Job Briefing

Line switching, workers' protection assurance, lock out/ tag out and other procedures can add lengthy time to a task that only takes a few minutes. Linemen need to remember that in line work, however, a line is not dead until it's tested and grounded. In job-site briefings, linemen must follow all the proper procedures and discuss all of the energy source controls to stay safe. It's that simple.

Lack of job planning plays a role in many serious injuries and fatalities. For that reason, it is crucial for line crews to always schedule a job briefing before work begins. Doing so will allow linemen to begin each job safe and finish strong.

20 Finding a MAP for Safety Committees Success!

Chances are that you have a safety team—most companies formed safety committees more than a decade ago. Chances are that you have many safety teams within your organization. In fact, you may have a safety team for each department. And, chances are those teams are greatly under performing.

The Behavior-based safety revolution of the last couple of decades introduced us to the Safety Committee, a team of individuals gathering at regular intervals to do something. The problem is that the team and management alike are not always sure what the team should be doing nor how it should get done. To be fair to our teams however, it's not all their fault. If most organizations are completely honest, our best safety teams happen by chance and are the exception not the rule.

The fact is that our workers like electricians, heavy equipment operators, maintenance personnel, skilled craft, lineman, etc., are trained to work with their hands. They have great physical skills like building, operating equipment, and making things work. They are adept at job planning, hazard recognition, and simply getting things done. That being said, these same men and women are out are not always 'great' at setting safety committee team goals. They are not necessarily skilled at drafting an effective agenda or leading an effective safety team. And why would they be? After all, their training and experience has been in other areas. That is why in order to have an even more effective safety committee, we need to offer a specific road map for success. We can use a process that focuses on the most important hazards and communicates awareness in a structured and systematic fashion. We need a clear path to safety committee success, and that path begins with a safety committee MAP: *Monthly Awareness Plan.*

Finding a MAP for Safety Committees Success!

The following is the outline for a monthly plan guaranteed to get your committee focused and get your organization results.

Begin the Drive

To begin the MAP process, the committee will brainstorm ideas using the following questions. Where are our workers taking shortcuts? What are the biggest hazards and exposures? What are we doing that gets us hurt? What safety rules are most important, and which ones are neglected? What does our previous peer observation data tell us? What training do we need this year?

Twenty minutes of brainstorming can produce a hundred or more possible themes—the focus is on your specific work place hazards. Once you have finished brainstorming, pick twelve from the larger list that you feel are the most important. Then, match a hazard to a month; this will give you safety themes for the year (one per month). You and the committee have the discretion to change a theme if needed but at least you have established your framework for the year.

Shifting Into Gear

For each theme the committee should develop a MAP, each MAP should include the following:

Two Safety Activities. A safety activity is something done outside of a safety meeting that brings awareness to safety and your monthly theme. It might be a note taped to a locker, stickers for hard hats or awareness items in trucks or at workstations etc.

An Involved Safety Meeting Activity, which is a safety meeting that involves the entire group. It can be a game, activity or exercise that gets everyone involved. For example, if your topic is traffic protection, have your people get in groups of five. Give each a piece of paper with a section of road. It could be an intersection, a street or four lane highway, but a different one for each group. Have each group draw the proper traffic protection on the sheet of paper then present it to the group. That's a much better way to review traffic protection than twenty minutes reading from the safety manual. Remember, if your workers hear it, they remember about 10 percent of the message, if they see it and hear it (video) they may retain about 25 percent of the message, but if they do it, studies show they retain as much as 75 percent of the message! (There are references that can help with this, such as my book, *ISMA—101 Ways to Get Your People Involved*.)

An Outside Safety Speaker. This is an individual from outside of the work group that can share his/her insights on the monthly theme. It might be a subject matter expert from the community (police officer, doctor), an individual from another part of the company, or another worker who can share a near miss story. Learning about the topic from an outside speaker keeps the message alive.

Distribute at least one type of safety educational material during the month; this is an item such as a quiz, cross word puzzle or other educational game. It can be used during a safety meeting to educate the group on the monthly theme. These can be easily found on the Internet or created using inexpensive 'word game' software.

Finally, hand out at least one Safety Awareness Item, or SAI. A SAI is a tangible item that is given to employees to bring continued awareness to the monthly safety theme. For example, a Crunch candy bar can be used to remind coworkers not to be caught in a 'line of fire' crunch. The point is to keep the awareness high and get the group talking, and a SAI can do just that.

Once each activity, speaker, involved safety meeting, etc., is chosen, assign responsibility for each activity to members of the committee. This ensures the task is completed on time.

In the end, to foster safety committee success, we must create involvement, teamwork, high energy, focus, communication around the most important hazards, and the drive to change behavior. The safety committee MAP process can do all of that, and more. The only real question is, are you ready to open the MAP?

21 The S.T.O.R.M. Model

One Approach to Effective Near Miss Reporting

In business and in safety, it is not about 'if' a storm will hit, instead it is about when a storm will hit, how many, how severe and how we will react when they do hit! Having worked more than a decade and a half in the utility industry, I saw a number of devastating storms, first hand. I'm talking about figurative storms such as downsizing, layoffs, financial crisis, etc. I also saw a number of Mother Nature delivered storms, the kind that uproot trees, tear off roofs and destroy power lines.

It is safe to say that the most dangerous work for utility workers is during storm restoration. After a storm, weather conditions are generally poor and electrical hazards extreme. Hundreds, if not thousands of workers from all different companies compress into a relatively small area to restore power. Workers, line-managers and safety staff alike work sixteen, eighteen and sometimes twenty-hour shifts with one goal, to restore power in a safe and efficient manner.

To help mitigate these extreme hazards and maintain a high level of safety awareness, crews will gather every morning to review hazards, work areas and assignments and then meet again in the evening to review the day. It's in these evening sessions that line workers will openly share near miss events. It's a free exchange motivated by a general intent to keep a fellow "brother or sister" from being injured by the same or similar exposure.

The process is supported by line-management and safety staff alike, it is free, workers are motivated by a sense of genuine caring and a desire to help, it's not full of process or forms and it is done in a timely manner. Yet, after a storm, most of this free sharing is lost as crews revert back to "normal" hazards and office politics.

So, what if we can capture this 'storm' mind-set into everyday work life—we can.

Near miss reporting is extremely important to every organization. A few months ago, I published an article called, "Finding the Smoking Gun—The Five Things That No One Will Talk About in Near Miss Reporting and Questions to Foster an Effective Reporting Environment." In that article, I pointed out the real-dollar cost of near miss incidents and listed five reasons that no one reports near miss events. The feedback from that article was tremendous with many readers challenging me to offer a solution in addition to the problem. To that end I offer the S.T.O.R.M. model for near miss reporting. It's taken from the free exchange of line workers during storms and shaped so that any and every organization can foster this approach within their respective group.

The S.T.O.R.M. Model

S: See the Near Miss—One of the most effective football coaches of all time, Vince Lombardi, used to say, "It's hard to be effective when you are confused!" That same spirit translates into near miss reporting. The fact is that our people have a hard time defining a near miss and have a hard time translating a field event to a near miss report. And, when there is uncertainty, people will withdraw into silence and status quo. Make it clear that any incident, 'strange happening' or vital piece of information can safety be shared at the daily briefing.

T: Tell the Group at the Next Daily Briefing—Set a routine of holding a daily briefing. These are typically short, effective safety awareness meetings and not a typical safety meeting. I suggest that these be held in a different location than your typical safety meetings. Since these are typically short, it's okay to stand in a circle versus sitting. Be prepared for the meeting. It's a good idea to begin each meeting with a meeting starter, such as a safety related news story, a quote or some others short safety related piece; if you need ideas check out my book, *Tailgate-101: Stories to Start Each Job Strong and Finish Safe!* After the meeting starter, invite the group to share any near miss, incident or safety incite from the previous day or shift. After this sharing, point out any specific hazards for the group such as weather, changing in field conditions, changes to a specific project, etc. Then, end the meeting and begin safe work.

O: Own to Eliminate—In this step, analysis each near miss that was reported and determine any and all follow up needed to eliminate the hazard; own it! In this step, it may be important to go back to the person who reported it and ask for help and support in making an action plan for future mitigation. While it seems that participation in the previous section is key, I feel that this is the section where management and safety leaders will earn long-term credibility and trust if and only if they are able to properly address and eliminate key hazards through near miss reporting.

R: Review/Remind—During the daily job briefings, it is important to review with the group follow up action steps taken as a result of incidents shared during the daily briefings. It is also important to remind the group of hazards reported and the safety rules and procedures needed to safety deal with such hazards.

M: Move On—Finally, move on. The process is designed so that there is very little or no 'red tape' or paper work; instead a free exchange with important follow up by management, safety leaders and workers involved, when needed. After an issue has quickly been resolved, move on. Keep the material fresh, and the group focused on immediate hazards and near miss events.

"The only ship in a storm," an anonymous quote reads, "is leadership!" Storms will come and storms will go. Near miss events left unattended however will continue until someone is finally injured and the situation addressed. Providing your group with the leadership that allows for swift and effective sharing of near miss events and an equally swift and effective mitigation of the hazards is key not only to credibility, trust and long-term safety success. Take advantage of the S.T.O.R.M. model, for our seas may not be smooth, they can be safe and injury free!

22 The Space Between
What's in Your Space?

The cell phone rang; being in a meeting, I ignored it. It immediately rang again and I stepped out of the room. It was the regional dispatcher. I can still remember his words, "Electrical contact Matt, we've got two men down."

Anyone who has ever climbed a pole or worn rubber gloves knows safety. We understand PPE, from a sticker on the hardhat to the steel toes in our boots. We are well trained; we know our work and appreciate the hazards associated with it. We begin as apprentices, able to slowly absorb the safety rules that apply to the job. And, we are physically fit, able to climb poles, lift cross-arms and if push comes to shove, hoist a 50 KVA tub with a couple of men and a set of blocks. Yet, day in and day out across our industry, we have men and women who get hurt and unfortunately even killed.

After several years in the field as a lineman, a foreman and safety professional, I wondered, we are skilled and knowledgeable to perform our work safely, so why do we still get hurt? What I discovered is that there is a gap; a space between what we know is right and the shortcut we sometimes take. Realizing this, I asked the next question, what's in it? You will never guess what I found!

Dilly-Dally—As linemen, we love our tools! But, lurking in the space between what we know is right and the shortcut we sometimes take is dilly-dally and it has to do with our tools. Dilly-dally is when we use tools wrong or don't take the time to get the right one. And don't tell me this doesn't happen. I remember when one guy was drilling a pole hole. The wind was blowing and he was catching wood chips in this face. He began to call for additional protection, which was in the truck bin, but he decided to use his hand as a shield instead. We spent the afternoon in the emergency room flushing his eye. Or one job, we parked our one-ton pickup truck and trailer. Next to the trailer was a piece of wood and we kicked it under the trailer, you know, a wheel chock. As we unloaded the back yard machine, the weight shifted and the whole rig went sliding down the hill. It sailed through a yard and rested against a tree. Apparently, it was a very expensive tree, owned by a prominent attorney. By the way, the real wheel chock was on the trailer, kicking the stick of wood under the wheel was just easier. Dilly-dally isn't all wood chips and band-aids. A few years back a crew was working in the yard and needed to take a bracket off a trailer. It was about 12 feet in the air and while they didn't have a ladder close; they did have keys to the forklift. Near by, was a cage resting on a wooden pallet. It was the kind raised and lowered by a crane. The cage on pallet was retrieved by the forklift and one of the men climbed in. He was raised to remove the bracket, who needs a ladder? All went well, until the man in the cage shifted his weight. The cage came off the pallet and the man with it. He suffered massive head trauma and was killed. His five-year-old son didn't care about dilly-dally, he only cared that his dad was gone—forever.

Bullet Proof—When I made foreman in a small, rural Missouri town, I supervised a couple of crews and two troublemen. The very first job I pulled up on, one of my troublemen, Larry, was in the bucket hooking up a service. Back then, there were only five safety rules that applied to this job; one had to wear a hard hat, safety glasses, harness, low voltage rubber gloves and put a wheel chock out because the boom was in the air. How many rules was Larry following? One, he was blind as a bat without glasses and did have prescription safety glasses. I called for him to come down and he refused. I called again. I had never seen Larry move so fast. He stowed the bucket and jumped out. He dashed over to where I was like a linebacker. I was in great shape, 6' 2" and 190 lbs., but Larry was in better shape for a sixty-something. His 6' 4" frame peered down through my eyes and he finally asked, "What?" I told him that he needed to wear a hard hat, harness, low voltage gloves and stick a chock under that wheel. He was steaming and he finally said, "Son, I've been doing this since before you were alive, I know what to do and when to do it."

In reality, following the rules is up to you—it's your call. And while I was able to work with Larry for some time and saw his habits change for the better, if you don't care, I don't either. But, understand the deal. In the space between what we know to be right, and the shortcut we sometimes take, is a bulletproof vest. We strap the vest on under our flame retardant clothing and claim, "I know what to do and when to do it." But, one day you will be running late. You will kiss your kids and hug your wife. You will grab the lunchbox but forget the bulletproof vest stuffed in the corner; by the way, my condolences to your family.

The Black Lung—I told the dispatcher that I'd be there as soon as I could. I left the meeting and peeled out of the parking lot. I had an 80-mile drive to the work site; how could this have happened?

The crew had been in nothing but hot work for the last three weeks. They were setting poles and laying out phases to reconductor a piece of line. It was September and breezy and Friday morning. The six man crew, with over 100 years experience between them, was going to work on this last pole and would be ready to pull wire on Monday. In all, this was a cake job for a Friday. It was so easy, setting a pole and laying out the phases, they didn't tailgate. It was so easy no one talked about the phase that was over the truck. Nothing would go wrong here, so the operator moved the boom to pick up the pole without a spotter. It was so simple, two men were getting material off the truck as the boom operator stuck the pole grabbers in the 12 kV phase while lifting the pole. "Electrical contact Matt, we've got two men down."

The black lung went primetime a few years ago when anti-cigarette advocates used it to convince teenagers not to smoke. The argument was simple; it said if one smoked their lungs would turn black, years later, after much pain and suffering, they would die. The reasoning was solid, an attempt to scare kids into not smoking. There was only one problem; it didn't work. The reason, people, young and old, didn't buy it. The truth is that we don't really think anything bad will happen to us. While the tobacco industry made the term 'black lung' a household word, the same scare tactic has been around the safety industry since workers began taking risks. We read graphic incident summaries. We preach that 'it' can happen. We show films with blood in an effort to scare apprentices. The problem, in that space between what we know is right and the shortcut we sometimes take is a black lung, the belief that it can't happen to us. It did happen to this crew; Frank and Stacey were down. Frank didn't make it . . . .

In conclusion let me leave you with the following thoughts. First, meet Joe. He was working in a substation and needed to remove a set of grounds. He couldn't break the ground loose with the shotgun stick so he climbed off of the ladder onto the top of a transformer. He was about 12 feet in the air and didn't think he needed fall protection; after all, he would only be there a second. With leather gloves, he used a screwdriver to break the ground lead free. When he removed the ground he was immediately shocked by induction voltage. Not enough to kill him, just enough to knock him off the transformer. In that instant, falling head first, he saw his son and wondered who would take him deer hunting. In that wink of an eye, he saw his daughter and questioned who would walk her down the isle. He hit the ground and all went black.

Safety is not about fluff or 'rah-rah' presentations or glossy posters. Instead, it comes down to a simple truth, a mighty secret and a harsh reality. The truth is that topping out as a journeymen means we are trained, skilled and knowledgeable to do our work safely. The mighty secret is that there is a space between; and success in safety is dependent on our ability to control what dwells in this space. Last, the reality is that when working in the space between we can lose it all. While Joe woke up, spent months in the hospital, and had several surgeries, he did recover. Many that you and I both know were not so lucky.

So what do we do with this space between? Avoid it like sour milk or the common cold—it's bad! There may be a number of ways to make this choice, but the one of the best my come from an old working foreman that I worked for as an apprentice. This old foreman was as strong as an ox and could work any of us young guys into the ground. He understood the space between, however. Each day we would stop to have a tailgate and discuss the job hazards. We would talk about our work, who would do what, and how to do it safely. Most importantly, at the end of many tailgates, he would look each man in the eye and say, "And one last thing, no one gets hurt today." In other words, stay out of that space! From one lineman to another, never forget that there is a space between what we know is right and the shortcut we sometimes take. What lives in your space?

23 Small Stuff Matters
So Everyone Finishes Safe!

Although you probably won't admit it to your family or your boss, there are probably a few rules that you don't follow. I don't care what your role is in the utilities and or the transmission and distribution world, (lineman, substation tech, meter tester), there are a few rules that you have probably deemed 'unnecessary' or 'irrelevant.' Am a right?

Before I talk about these rules, I wanted to first talk Navy SEALS. SEALs, which stands for Sea, Air, and Land, were organized in 1962. Their mission was to be the best trained fighting force in the world, and utilized teamwork to move against a target in which a larger force could not approach undetected. "While it is imperative the student meets the standards set before him," said Intelligence Specialist 2nd Class Matthew Peterson, SEAL instructor. "We look for the individual who possesses the ability to perform safely and effectively under stressful conditions. Ultimately, we are seeking a candidate that we can entrust with the life of a fellow Frogman."

In order to be called a Navy SEAL, one must complete the 25-week BUD/S (Basic Underwater Demolition School). The dropout rate of this school can be as high as 70–80 percent. The training is divided into three phases. The first eight-week phase is known as the physical conditioning phase, and places a strong emphasis on running, swimming, navigating the obstacle course, and basic water and lifesaving skills. In this phase of training, Sailor's will take part in "surf conditioning," an exercise that develops teamwork among the trainees as they lie in a line with arms connected while the cold California surf washes over them.

In his book *Leadership Lessons of the Navy SEALS,* Jeff Cannon, a former Navy SEAL, shared this story of his first phase of training. He said that when he was in BUD/S training, his class had just finished 'drown proof' training. That's when you stay in the water for hours learning not to drown. Finally the instructor called them to shore, they were exhausted. The instructor said it was 'dinner time' but before they ate, they needed to give him five miles on the beach. The only instruction was; 'you have thirty minutes and everyone finishes on time." Exhausted, they set out running. Some did finish within the half hour, others didn't, the instructor wasn't happy. As the last of the trainees crossed the finish line, the instructor said, "You obviously didn't understand me, I said everyone finishes on time. Now, do it again."

Our jobs are complex, and there are hundreds of rules and procedures. Over time, there have probably been some rules that have simply dropped out of habit or off your radar. Let's look at a few.

Do you use hotsticks? Utility line workers use them all of the time in switching and live line work. The last time you used a hotstick; did you wipe it down before use and perform a visual inspection? OSHA, in 1910.269(j)(2)(i), says, "Each live-line tool shall be wiped clean and visually inspected for defects before use each day."

Rubber gloves may be the most important tool that a transmission and distribution line worker uses. Ones life literally depends on the care and integrity of that glove. But do we perform a visual inspection and a field air test on our rubber gloves before *every* use? Sure, we test them when they are new and probably before a big job, but probably not before every use, every time.

Steel-toed boots? I understand rules vary across companies and regions based on job responsibilities and exposure, but I also understand that many companies require steel-toed boots at all times! Several years ago, a line apprentice had a nice pair of high-end steel-toed line boots. After a couple of years of wear, he sent them back to have them rebuilt. Instead of buying another pair of steel-toed boots to wear when his good ones were out of service, he chose to wear an older pair without steel toes. After all, it was only for a couple of weeks, what could happen? Five days later, he was on a crew working on some underground manholes and he dropped a manhole cover on his foot. Without steel-toed boots to protect his feet, he broke three toes and needed pins to repair them.

Slings? Most of the time, we grab a sling from a truck and use it. After all, if it's on the truck, it must be okay. Yet, OSHA, 1910.184(d) tells us to inspect a sling and all of its fastenings and attachments before use, every time!

How about wheel chocks? Many companies require, as do manufacturers, that if the boom is in the air, wheel chocks must be out. What is your habit and do you even have a chock on your truck?

Later in his book, Jeff Cannon talked about Navy SEAL dress inspections. He said that after graduating from SEAL training, SEALs may be deployed in locations were they need to blend in, grow longer hair, a beard, and maybe dress sloppy. Yet, when dress inspections were ordered, each SEAL took time to show up for the inspections without missing a detail. One may think that once graduating from training, and working with high explosives in adverse conditions, that a SEAL won't care about a crease here and polish there, yet it is actually just the opposite. Each SEAL understands that if he can't be entrusted with the small stuff, like polish on boots, he won't take care of the big stuff when the pressure is on.

Our work is the same. We may sometimes think that a missed wheel chock here or skipped inspection there is not a big deal. We might reason that we just inspected our gloves at the last job, no reason to do it here. But, if we are going to follow the big rules and entrusted as our brother and sister's keeper, we have to focus on the small stuff... each time... every time.

By the way, Jeff Cannon says that when they ran the second five miles, each person on his team finished on time. Some were pushed, some were carried but everyone finished. Today, let's make sure that we pay attention to the small stuff, so that everyone on our team can finish safe and go home at the end of the day.

Source:
Cannon, Jeff, *Leadership Lessons of the Navy SEALS: Battle-Tested Strategies for Creating Successful Organizations and Inspiring Extraordinary Results*, McGraw-Hill, 2002.

24 Safety's Broken Windows

Why Wheel Chocks and Steel-Toed Shoes Really Matter!

In the mid 1980s, New York had a crime problem. While they weren't the only major city that was seemingly losing the crime battle, they took a very unique approach in trying to solve it. What did they do to solve a crime epidemic? They started with graffiti on the subway!

In 1985, the New York City Transit Authority was at a crossroads. The system was aging and crime on subways was seemingly out of control. The Transit Authority hired a new director, David Gunn, to solve the problem. Gunn inherited a list of complex issues such as budgets, crime on subways, an aging system, questionable management structure, etc. Gunn would lead a multi-billon-dollar effort to rebuild this system and many experts were telling him to start with system reliability and violent crimes. Instead, he ignored the advice of the consultants and decided to try an unusual tact for improvement to turn around this seemingly broken system, graffiti. He set the goal to eliminate graffiti and vandalism from the subway system.

He began, car-by-car and train-by-train to retake the subway system. He began with the Number 7 train that connects Queens to mid-town Manhattan, taking the train out of service and removing all vandalism and graffiti. He established a rule that once a car was cleaned and put back into service any new graffiti would be removed immediately. Cleaning stations were established and as a train finished its run, it would be pulled into the cleaning station and any new graffiti removed. If it could not be removed, the train would be pulled from service until all signs of the vandalism were gone. This effort, which began in 1985, took until 1990 to complete.

What Utility Safety Leaders Do

In 1990, the Transit Authority named William Bratton to head the Transit Authority Police. In this role, Bratton, a military veteran and life-long law enforcement officer, took another unique position in solving the crime problem. With felonies and violent crimes on the subway system at record levels, Bratton did not target subway drug use or violent offenders. Instead, Bratton, like Gunn, defied experts and set a goal of cracking down on fare beating. You know, fare beaters are those who jump over the turnstiles to save the $1.25 subway fare.

Bratton wanted to put an end to the 170,000 people who were entering the system without paying. He began to set up undercover teams who worked to catch fare beaters. Once nabbed, these fare beaters were cuffed and placed in a holding area within the subway terminal. It was a location that everyone entering and leaving the subway could see. Once the holding cell was full, all detained would be hauled to a booking bus, processed, finger printed, and a criminal background check performed. Like Gunn with graffiti, Bratton wanted to send a message to fare beaters that this behavior would no longer be tolerated.

The broken windows theory was articulated by criminologist James Q. Wilson and George L. Kelling in 1982. Their theory reasoned that if a building has a broken window that goes unrepaired for a long period of time, vandals will recognize the broken window as a signal that no one cares. Since no one cares, vandals will be inclined to break more windows. Broken windows will quickly lead to other forms of vandalism, then breaking into the building, which will lead to more crimes both in the building and the surrounding areas.

There has been a strong trend over the last half-decade to crack down on the safety equivalent of violent offenders. Thousands of organizations and industries over the last few years have established 'rules to live by' or 'cardinal sin' lists. These are lists of safety violations so egregious that if violated, one could experience a life-changing event and even death. Depending on the industry, this can include failure to wear fall protection or effectively tie off, failure to effectively guard against electrical hazards or improper confined space entry. If employees are 'caught' violating one of these 'felony'-type safety rules, they are immediately suspended.

"Don't be afraid to give your best to what seemingly are small jobs," motivator and trainer Dale Carnegie would say, "every time you conquer one it makes you that much stronger. If you do the little jobs well, the big ones tend to take care of themselves." While I don't disagree with cardinal rules, I think it identifies a trend of only looking toward 'the big one.' This is done often at the expense of wheel chocks and earplugs. "The graffiti was symbolic of the collapse of the system," Gunn later said. "When you look at the process of rebuilding the system and moral, you had to win the battle of graffiti." For Gunn, graffiti was a broken window. The idea was to communicate a very clear message to vandals—the windows were no longer broken! For Bratton, the broken window was fare beating. It represented a signal that led to more violent crimes. Once one or two people were fare beating, others would think that if someone else wasn't paying then they weren't going to pay either. If one window is broke, then another one won't hurt.

Keep the focus on 'big picture' and widen that picture to identify safety's broken windows. Where is your graffiti and fare beating? What signal is sent when employees turn the other check on very small and simple rules such as wheel chocks, earplugs or having the right number of buttons buttoned on a flame retardant shirt? These, and other seemingly 'insignificant' safety practices are our broken windows. This is a signal from management to the employees of what is important. In learning from Gunn and Bratton, we understand that a vigilant focus on the single broken window might actually net the results for which we have been looking. In the end, Bratton found that one out of seven arrested for fare beating had an outstanding warrant. He found that one out of 20 were carrying an illegal weapon. In safety, those who violate the smaller, seemingly less 'dangerous' safety rules are giving a signal of a broken window. A window in need of repair before it's too late and greater damage is done. Erase the graffiti today!

25 Safety Is NOT the Right Thing to Do!

Since we can remember, we are told to do the right thing. As children we are told that being nice to our brother or sister is the right thing to do. In grade school we are directed to be kind to all children on the playground or to that one child who is having a tough time. As teenagers going out with friends on a Friday night, we are reminded to 'do the right thing.' In fact, many utility leaders have now adopted this slogan for all parts of their business, especially safety—safety is the right thing to do!

I too had bought off on this slogan, 'safety is the right thing to do' until recently. A serious incident made me reconsider this approach. The incident involved a three-person utility electrical line crew. They were working to change out a cross arm on a three phase, 12 kV circuit. One of the lineman got too high and put his shoulder in the middle phase as he was moving the neutral. He wasn't wearing rubber gloves nor was the middle phase adequately covered with protective rubber cover. His injuries are serious and significant, but lucky for him and his family, he will live.

If you asked this crew they would tell you that they were absolutely doing 'the right thing.' After all, 'the right thing isn't cut and dry. In this case the work needed to be done quickly. The crew was on over time. Other customers needed to have their power restored. Sure, they cut some corners, some major corners, failing to follow even the most basic of rules, but if you ask them, they were doing the right thing.

From this event and others I have drawn a new conclusion. What we term the 'right thing' can be confusing. It can lead to shades of gray. It can mean that sometimes cutting a corner appears to actually be the 'right thing.' So, the next time you hear someone say, "Safety—it's the right thing to do," stop and correct them. Safety is NOT the right thing to do, "Safety is doing the right thing." The right thing in utility work is cut and dry. It's measured by rules, planning, actions, hazard evaluations, energy source controls, and more. Below are four areas where safety is doing the right thing.

Follow All Rules—If we went into surgery and the doctor told us that he was only going to follow about four out of every five rules or guidelines for the surgery, what would we do? We'd find another surgeon! In the spring of 2003 I was doing a safety audit on a seven person line crew. I saw some risk and stopped the job. I called everyone together and told the crew they needed to change the way they were performing a certain task. I was told that they didn't need to change, that they were doing the work the way they had always done it (doing the right thing). We pulled out the safety manual and I referenced the section clearly telling them to do it another way. They acted astonished, asking how long that this particular rule had been in the manual. I said since the last safety manual revision—the revision date was 1983! Doing what is right means we know and follow all rules. We wouldn't excuse a doctor, lawyer, teacher or engineer for not knowing a rule; it's no different with our profession!

Wear Rubber Gloves and Ground Lines—Not too long ago, 'the right thing to do' was to de-energize a 12 kV or 4 kV, line then work it as dead, no rubber gloves and no grounds. It was a firm shade of gray and many distribution line workers chose this path. It was acceptable and it was clearly the right thing to do. But, doing what is right would mean there is no shade of gray—a line is either hot (energized) and all of the gloving and cover up rules apply, or it is de-energized and grounded. That's the right thing.

Plan Your Work and Work Your Plan—I find it interesting that within the OSHA standard for electrical line work, that OSHA specifically requires job planning. Not only do they require job planning, they tell us what to discuss. We are clearly instructed to pull the entire crew before any work begins, and we are clearly told what to talk about (HS2EP): First, hazards associated with the job. Next, safety rules that we'll need to follow. Special precautions, like all of those hidden traps or unique job situations, is third. Energy source controls is the fourth item. PPE is the final item. And, OSHA says that if the job or conditions change, you'll need to hold another job briefing to discuss. Planning this well is doing what is right.

Report Near Misses—I love storm work. The reason I do is that I learn so much—it's utility line work on steroids! One of the best parts of storm work is the morning brief. For most utilities, each morning of a storm, before everyone hits the field, they gather to discuss the number of outages, goals for the day and any near misses from the previous day. And, in storm work, everyone shares, because hazards are extreme and everyone is helping out so that no one gets hurt. But, when the storm is over, our line workers will return to their 'routine.' Much of which includes not sharing near misses. Doing the what is right is to make every morning like storm report. Take five minutes, pull your crews together to review the day, any changing conditions, the goals for the day and ask for near miss reports. Your people will do what's right!

So remember, there is a difference between 'safety is the right thing to do' and simply doing the right thing. And, that difference can be the difference between following a rule or taking a risk—in our business, the difference between life and serious injury or death. Do what is right; follow the rules, wear rubber gloves and ground lines, plan your work, and report near misses.

26 Five Keys to Leading Lineman Safety

Getting Respect and Getting Results

Just because it's a 'team' doesn't mean that it is led or coached the same way. One wouldn't inspire an NFL Championship team the same way as an Olympic medal winning Synchronized Swim team. I mean both are champions, yet both would be motivated much differently. One wouldn't coach the best information technology (IT) team the same as the best furniture moving company. Again, both may be recognized for excellence, but different leadership methods would be used for each. The point is simple, to be effective we have to coach to our audience; in leadership one size doesn't fit all. So, to be successful as a leader in line worker and utility safety, we must coach to our audience.

Line and utility work is unique. It's tough and grueling work all hours of the day and night. It is missing holidays and family trips due to storms and equipment failure and 'cars hitting poles.' It's holding energized wires and climbing poles, digging holes and shooting trouble. It's dangerous work that needs effective safety leadership. Leadership in this field must be earned, not by what is said, but instead by what is done consistently over time. Below are five traits to leading utility safety, gaining respect and earning results. After all, the men and women of this trade are watching—what do they see you do?

Care—Randy was a tough, hard to get to know, son of a gun. He had more than twenty-five years of experience and didn't get along the best with the local management. I knew Randy and had actually worked with him when I was an apprentice. I was now the safety supervisor for the area. In this roll I was responsible for nearly 400 linemen, substation workers and gas employees in out state Missouri. Knowing that I couldn't see everyone in the course of a month, I decided to show them I cared in a different way. I got with Human Resources and got everyone's birthday. Then, I wrote everyone a birthday note. For Randy I said something like, "Happy Birthday, I really liked working with you back in the day. I always liked your funny stories. Work safe, Matt." I sent it and forgot about it. About five years later I happened to be in Randy's show up location and there on his locker with the pictures of his wife and kids was a faded piece of paper. I recognized it immediately; it was my note to him.

Family counselor and author Josh McDowell wrote, "Rules without relationship causes rebellion." If I needed to talk safety or enforce a rule with Randy, I'm confident he would have listened because he knew that I cared. Care, so that when you need to talk about safety, your audience will be more than willing to listen.

Own It—Have you ever rented a car? If you answered 'yes', did you wash that rental car? Unless your some kind of 'sick' you absolutely did not wash it. It's a rental—you don't own it. I have been in dozens, if not hundreds, of meetings were a safety question was asked. The supervisor, safety committee chair, manager, responded to the question by saying, "Yeah, the safety department says we have to do it that way." Either you expect it from your team, or it's not important and blaming the safety department for a rule isn't owning it; it's the easy way out. With this response, your team knows that the rule in question is not important to you, and they question in what other areas you take 'the easy way out.' In safety, do you own it?

Know Why—There is no doubt that utility work is complicated. We ground, wear rubber gloves, use load break tools, stick primary and the list goes on and on. And, with every tool, rule and procedure, there is reason why. One of the best leaders I have ever worked around was Joe. After serving our country in Vietnam, Joe got on with the local utility. Over time he was promoted and found a home as an apprentice trainer. In the position he became a leader in the entire company for one simple reason, he knew why. If you wanted to know why rubber blankets were tested but hard cover was not, Joe knew. If you wanted to know why OSHA requires a hot stick to be wiped down before each use, or how the minimum approach distance was determined, Joe could honestly tell you. Linemen, more times than not, will follow the rules, if they only know why. If you can't remember 'why' then find the Joe in your organization so that when your people ask, you can get a quick answer.

Relate—I was blessed to work around a utility gas journeyman who always said, "It will happen. What matters is where you are when it happens." His point was simple, mechanical equipment does fail and people make mistakes—thus, bad things will happen. But, if you have taken the time to plan your work, follow safe work practices and wear the proper PPE, when 'it' happens, you will be okay.

"We work to live, not live to work." To motivate utility workers to be prepared when 'it' happens, this gas utility worker was a master at relating. He would ask one very simple question—why are you here. Once you get through the first few layers of superficial answers, people are at work for family, to put kids through college, for a new bass boat, for retirement, to travel. It doesn't matter why someone is at work; it only matters that he or she understands that none of these dreams and goals happen if they are not prepared when 'it' happens. Relate work to life consistently over time and you will find yourself leading safe work.

Jump In—Tony was an award winning safety professional. He was recognized by the safety profession as 'the safety professional of the year' in a large mid-west region. With nearly twenty years of experience in construction and manufacturing safety, Tony changed fields and came on board the safety staff of a major utility. The only problem, Tony didn't know a wheel check from a fused cutout. About three weeks after Tony started, Hurricane Dennis struck the gulf coast. To restore power, lineman from all across the country blanketed the area, and Tony was asked to accompany about fifty linemen on a three week tour. He was nervous to go, after all he didn't know many of the rules, but he went none-the-less. Tony spent every waking hour in the field with the linemen. With safety manual in hand, he watched and asked questions. He put on a harness and went up in a bucket. He influenced safe choice not because he was the authority, but because he was willing to roll up his sleeves, be a part of the team and ask questions. Many people work in safety and management roles directly supporting utilities may have never performed the work. And, in these instances there is often a high hurdle to gain the respect and credibility of the 'team,' since you didn't actually 'play the sport.' In these cases, lead like Tony—and jump in. You will be amazed at the results.

In the end, lead utility workers the way that they need to be lead—and you will find new and amazing results.

27 Is Your Culture Killing You?

How to Move from a Culture of Honor to a Culture of Safety

Did you know that if you are a male less than 40 years old living in a rural part of a southern state you are nearly 20 percent more likely to be killed by an accident; compared to your peers from Northern states? And, this accident would stem from risky behavior. "The leading causes of death in people ages 1 through 44 are unintentional injuries, so we're looking at a substantial public health problem," said Dr. Paul Ragan, associate professor of psychiatry at Vanderbilt University Medical Center in Nashville, Tennessee. Centers for Disease Control and Prevention estimates more than 7,000 accidental deaths each year are linked to this 'risky behavior.' This 'substantial public health problem' is now being called the Culture of Honor.

"Cultures of Honor," writes Malcolm Gladwell in his book *Outliers: The Story of Success*, "tend to take root in highlands and other marginally fertile areas such as Sicily and other mountainous Basque regions of Spain. If you live on some rocky mountainside, the explanation goes, you can't farm. You probably raise goats or sheep, and the kind of culture that grows up around being a herdsman is very different from the culture that grows up around growing crops. The survival of a farmer depends on the cooperation of others in the community. But a herdsman is off by himself. So he has to be aggressive: he has to make it clear, through his words and deeds, that he is not weak. He has to be able to fight in response to even the slightest challenge to his reputation—and that's what the culture of honor means."

So why does this affect the southern states more than other states? Researchers believe it has to do with where the original inhabitants of the south came from. The southern areas of the United States were primarily settled by people coming from the most ferocious Culture of Honor—the areas of Scotland and Ireland. In fact, University of Michigan researchers tested the concept of 'the culture of honor' and found that the deciding factor on how aggressively an 18- to 20-year-old college student reacted to a threat was where they were from. The conclusion, cultural legacies are powerful forces!

Today, our organizations have cultural legacies, deep roots and long lives—and maybe none greater than in the utility and electric distribution industry. We are a proud group. We work in a hazardous field, and we have a strong historic culture. But, it is time to ask one basic question, do we have a 'culture of honor,' accepting risky behavior, or a culture of safety? Below are four questions to consider when you make that evaluation.

Do Small Things Matter?—For southern men, who have a one in five chance to be killed in 'risky behavior, sometimes those risks seem small. It's not wearing a helmet on the motorcycle or refusing to buckle the seat belt. In safety, small stuff matters. A number of years ago I was the lead on a work place fatality. It was a utility case where a boom touched a 12,000 volt line and one of the workers on the ground was killed. I arrived on location about two hours after the event occurred. In looking at the scene it bothered me that one simple detail was overlooked by the crew—the wheel chock. The utility preached that when booms were in the air, chocks needed to be down. But this crew overlooked a very small detail, the wheel chock, and they also failed to properly control a deadly hazard, the high voltage line. Are your people taking 'small' risks? What is that attitude toward wheel chocks, ear plugs, safety glasses, gloves, hard hats, seat belts? Failing to strongly adhere to these small things could mean you have a culture comfortable with risks —and not safety.

Is Planning Important?—When I was an electrical lineman apprentice, I had one old line foreman who would take extra time to plan. After a thorough job briefing involving he entire crew, he notoriously ended the planning session with, "And remember, nobody gets hurt today." Contrast this with what some other crew leaders did; park the trucks and went to work. After all, everyone knows what to do; there is no need to talk about it! One is a classic culture of safety and the other a culture of honor—which one is your organization?

Is Cowboying Allowed?—Have you ever heard these before? "He knows what he's doing, he's a journeyman," or, "Don't worry, he does it this way all of the time." What about this, "This one doesn't bother me, you should have seen what he did yesterday." If you hear comments like this, then you are in the middle of a culture of honor not safety. Enough said, cowboy.

Is Your Workgroup a Community?—"The family that eats together stays together," so the old saying reads, but I always say, "The work group that eats together is safe together." There is more to these sayings than just food. In today's busy work life, are we taking the time to build community? In farming areas community is key and neighbors not only help each other, they depend on each other. They are connected by a sense of obligation—to support and be supported. When workers feel connected, (community), they are more likely to speak up when something is unsafe, to share a near miss report and volunteer for that safety team. Is your work group a community? When is the last time you shared a meal?

The culture of honor is anything but 'honor' when it comes to at-risk behavior. Taking a shortcut doesn't honor the individual who is taking the shortcut nor his friends and family. Yet, like many cultures, they have just evolved. The opportunity that we have today is to intentionally start over—we can point our culture in a new direction. Leave the herd today and begin that journey to a culture of safety.

Sources:
Caroll, Kim, *"Honor Culture" Linked to Accidental Deaths*, August 15, 2011, American Broadcast Company.

Gladwell, Malcolm, *Outliers: The Story of Success*, Little, Brown and Company, 2008.

28 Is Training All Wrong?

What Leaders Recognize about Training

Years ago, I was a safety professional in charge of a large geographic area in out state Missouri. I was tasked with supporting nearly 400 linemen, substation technicians and natural gas pipefitters/equipment operators. My role included safety awareness, safety committee facilitation, reporting to senior management and a host of other responsibilities. I was also tasked with a large number of training models.

In the electric utility industry, OSHA demands a plan before any work begins. OSHA mandates a five step process to job planning. First, all participants on a job must meet before work begins and OSHA outlines the specific topics that must be covered before work begins. Just in case you are interested, those topics are; Hazards associated with the job, Safe work practices that apply to the work, Special precautions, asking the question, what is different about this work, or what are the hidden hazards that can get me hurt. Next, OSHA requires a thorough discussion of Energy Source control and finishes with the PPE required for the work. Or, as I like to say, HS2EP. I noted that on my many field observations that crews were not properly planning. And, I noted on a number of incident reports that lack of planning was contributing to injuries. I did what any proactive safety professional would have done—I put together a training program.

Working with another safety professional, we designed a great, (if I don't say so myself), three hour interactive program. We delivered the training and reinforced the message with cards for dashboards and billfolds. As a matter of fact, I still carry my card nearly ten years later! About three months after all training was complete, I was in speaking to one of these groups who, just 90 days prior, had this wonderful training, and I decided to give them a quiz. To my surprise and astonishment, it took several minutes before this group could recite the five key elements of a job briefing, or HS2EP. Alarmed, I thought it might be a fluke. I repeated the 'quiz' across much of my area only to find it was not a fluke. Most groups struggled to recall those key elements. Many were not using the cards. While job briefings may have been happening, my fear was that old habits had returned or more likely, never went away. And, due to a lack of planning hazards were being missed. I was troubled, why didn't the training stick?!

After much analysis and scrutiny, I believe the training failed because I finally realized I was not the trainer! Sure, I was the guy in front of the work group facilitating the three-hour session but I was not the trainer. Sure, I was the safety professional guiding the material, advancing the slides, but I was not the trainer. Sure, I instructed the work group on the right way to hold a job briefing, but I was not the trainer. What I failed to realize is that to change a work practice like this, training doesn't last three hours. It lasts 30 days! And, the first line supervisor is the trainer, not me!

Making It Stick—In order to make this type of training stick, safety professionals must do more than offer a three hour training. We must offer a three hour and thirty day training session. Before the initial training we need to:

▶ Get approval from managers that the training is needed.

▶ Be clear about and communicate well expectations follow up responsibilities of each supervisor. Meaning, that supervisors are the trainers and must do specific things after the initial session to form a habit.

▶ Next, provide the training to all supervisors, giving them the training and the set of expectations as follow up to the training.

▶ Provide resources to supervisors to help them be successful.

▶ Provide some specific evaluation process at 30 days, 90 days, and one year, to make sure the training did 'stick' and a new habit is in place.

Is Training All Wrong?

The New Skill—Over the last decade, many organizations have moved to get supervisors out from behind the desk and into the field. The intent is to have supervisors coaching our field or floor employees and offering feedback. The key skill set for supervisors is the ability to observe work and give timely feedback to correct at-risk behaviors and to reinforce positive and safe behaviors. Yet, the more we learn about key drivers to incidents, the more we understand that this key skill set is only half the battle and will not propel your organization to the level of safety success Let me explain the 'other half.'

A True 'Preventable' Tragedy—The cell phone rang; being in a meeting, I ignored it. It immediately rang again and I stepped out of the room. It was the regional dispatcher. I can still remember his words, "Electrical contact Matt, we've got two men down."

I told the dispatcher that I'd be there as soon as I could. I left the meeting and peeled out of the parking lot. I had an 80-mile drive to the work site; how could this have happened?

A study entitled "The Peer Principle" by *Bloomberg Businessweek* published in May 2010 stated, "In the area of safety, our study found that 93 percent of employees say they see urgent risks to life and limb, and yet less than one-fourth of those who see concerns speak up about them. Rather, they wait for bosses or others to take action."

Once on site, I found that the crew had been setting poles and laying out phases to reconductor three mile section of line. The six man crew, with over 100 years of experience between them, was going to work one last pole then go home for the weekend. Given the experience of the crew and the fact that this job was normally done with three men, not six, it was a cake job for a Friday.

The only major hazard on the job was a 12,470 volt phase-to-phase overhead line. The crew knowingly positioned their truck under the line to avoid setting up on a busy road. Putting the truck there, under the only hazard on the job that could quickly end one's life, one would think they would have stopped and discussed this hazard; or, placed a spotter designated to watch the boom, making sure it stays out of the minimum approach distance, or ground the truck or cover the lines. They did none of these things, remember, "less than one-fourth of those who see concerns speak up about them." Shortly after starting work, the boom contacted the overhead line as the men were pulling material off the truck. Both receive an electrical contact. One man died. A family is without a father. All five men, who were on the crew, will live with this memory forever.

Recently, researchers were studying organizations asking why there were differences in safety records. "We found," researchers later wrote, "That on the surface, the best and the rest looked quite similar. All were fastidious in keeping up with signage, inspections, compliance training, and enforcing safety policies. But we kept hearing unusual language in our interviews with the true standouts. It wasn't until we interviewed and surveyed 1,600 safety directors, managers, and employees that we realized we weren't really getting it."

In the end, researchers found that accountability was the key element to outstanding safety performance. But, it wasn't supervisors holding workers accountable that propelled organizations to the next level. Instead, it was workers holding each other accountable. "Since accountability appeared to be the key to safety as well as the full trove of corporate performance treasures, we then explored what made accountability tick in the leading teams and companies.

Remarkably, cultures of accountability had little to do with bosses. Rather, it was all about peers." Organizations with cultures of peers coaching peers found remarkable success—and not just in safety. "Those supervisors and managers with the strongest safety records were five times more likely to be ranked in the top 20 percent of their peers in every other area of performance. They were 500 percent more likely to be stars in productivity and efficiency and employee satisfaction and quality, etc."

This research should change the way we do business. Today, in most organizations, supervisors perform a prescribed number of job observations each week or month. And, we train these supervisors to observe crew work, look for safety rules and procedures that are not being followed, and then coach crews to perform work in a safe manner. We do NOT coach supervisors to access the level of feedback or accountability given on the crew, then coach the crew in techniques to improve that feedback. See the difference?

Supervisors should have a keen eye on the safety rules being followed, or not, on a job. But, the 'other half', and arguably the more most important piece for long term success, should not be to record these job audits, coach then move on. Instead, their primary focus should be on the level of coaching and feedback the crew is offering each other. The first is catching fish for the crew, the latter is teaching the crew to fish, and if the crew on this day had been taught to fish coach each other, then they would have taken steps to eliminate the hazard, saving a life in the process.

Making It Stick

- Revise supervisor observation sheets to reflect coaching and feedback.
- Offer training to supervisors so they know what coaching and feedback skills are effective, which ones to encourage and when crew communication is not up to par.
- Coach the coaches by having safety professionals or others who are skilled in the area of observing coaching and feedback to accompany supervisors. The purpose of this is to give supervisors additional training in this important area.

Source:
Grenny, Joseph, "The Peer Principle," *Bloomberg Businessweek—The Influential Leader,* May 2010.

29 How are Your Executive Safety Skills?

My daughter, has moderate to severe Attention Deficient Disorder (ADD) and Sensory Integration Disorder. My wife, and I saw signs early—we weren't sure what these signs were, but we saw signs. Because of these 'signs,' from the time she was a toddler, she is ten years old now, we built a team around her and around us. This team included doctors and professionals of all types who could give us advice and guidance so that we could help her succeed.

On a recent visit to one of these professionals we were discussing homework. Our daughter's third grade teacher expects all of the students to copy tasks and homework assignments into a day planner. If the assignment is completed throughout the day, then it is to be 'x-ed' out in the planner. If it is not completed, then it is a homework assignment and will need to be completed that evening. My wife and I were explaining to the professional that our daughter didn't seem to have the organizational skills to consistently copy the assignment from the board, mark it off her day planner if it was completed, or if not bring home the books and papers needed for homework. It sounds simple to us as adults, but she was struggling with the concept.

The professional, who was listening carefully, finally smiled and said, "Don't worry! Asking any third grader, let alone one who has added challenges of ADD, to do what is being asked is very difficult. Most third graders do not have the executive skills to carry out this task." He explained, "This is something that I would see develop consistently in a fifth or sixth grader."

Since this visit, we have used the term 'executive skills' a lot. It helps us understand that our daughter, and all children, develop key skills over time, each at their own pace. The concept of 'executive skills' however is not limited to the development of children. It applies to all sorts of development, especially utility work and safety. Our workers are all at different stages of development and in different stages in life. We have new employees and apprentice workers, skilled journeyman and wise foreman and crew leaders. Each stage is still in development and each stage is still working on his or her executive safety skills. The following are some key executive skills to keep us safe!

Learning from the Mistakes of Others—A rule of the trade is to 'never make the same mistake twice.' Yet, we spend a lot of time in safety meetings reviewing near miss and injury reports—are we learning from the mistakes of others? How often do we change our behavior because of the mistakes of others? Learning and changing because of what others have experienced is a key executive safety skill!

Speaking Out—Recently I took a friend to pick up his car at a dealership. His car had been in the garage and was now ready. I was a decent acquaintance of the owner of the dealership, so I went in to see if he was around. The walk from my car to the owner's office was very short but on that walk no less than three people stopped me to ask if I needed help or assistance. One of these people seemed to be the maintenance supervisor, not related to sales. The owner wasn't there, but the next time I saw him I told him that his customer service is outstanding—relaying the experience of his team being so helpful. He said that years ago he had implemented the ten-foot rule. That means no matter who you are, from the top sales person to the janitor, if a customer is within ten feet, you stop and ask if they need help. And, from the perspective of a recent customer, it is impressive!

In the last couple of years, so much has been made in the safety arena about coaching and feedback or simply speaking out when it comes to safety. I have learned that this is an executive skill that everyone can learn. Whether it is speaking out to a customer, in the case of the car dealership, a co-worker regarding safety, it is an executive skill that makes a noticeable difference.

Brake—Having ADD is creative energy all of the time, a fun light-hearted spirit and intense focus on one thing now, and another a split second later. ADD kids are bright and fun, but one executive skill that is a must is the concept of a brake.

When our daughter is 'fast' or off track, we encourage her to 'brake'. That allows her to stop and refocus on what is most important at the time. Using ones brake when gloving primary, flagging traffic or performing other safety sensitive jobs is very important. It allows us to stop, if only for a few seconds, check all of the hazards and make sure we are getting it done the right way. After you have braked and checked your surroundings, you can safely continue. Braking is an executive skill that must be learned. Stopping a job or task to reevaluate the hazards and any changing conditions is a key to safety, and a key skill to master.

Thorough Planning—When I was an apprentice electrical line worker in Kirksville, Missouri I would be assigned from crew to crew. I recall one old line foreman who would always take the time to carefully review the job before anyone started work. As the new guy, they gave me time to ask questions, and I always had dozens of questions. As we ended the job briefing and went to work he would always say, "And remember, nobody gets hurt today!" Years later, as a safety professional for a utility, we found that failure to properly plan was a leading cause or a contributing factor to each injury. And the more severe the incident the more the lack of job planning played a significant role in contributing to the incident.

The point is; planning is an executive skill that must be mastered. Many utilities, construction companies, and contractors are now using job planning forms to make this more effective. No matter how you get results, job planning is an executive skill that you and your organization must have.

We're Getting Older—In our nearly twenty years of marriage, my wife and I have moved, on average, once every other year. We recently decided to do it again. Last summer we purchased a lot in a subdivision, and several weeks ago, we sold our house. The plan is that we are moving into a rental house, then we will start construction on our home—these things always sound better when planning than in practice! On moving day, we had a number of friends and several hand trucks (dollies) and we went to work. About half way through, someone stopped and said, "Man, my back hurts, we should have all stopped to stretch."

We are all getting older; this is a key point to remember for those who earn a living with their hands and the sweat of their brow. Getting older means buying into a stretch and flex program, whether offered by your employer or done on your own. It means eating right and getting the right amount of rest. What we could get by with in our twenties can leave lasting pain in our forties—if you don't believe me, just ask a few of the weekend warriors who helped me move.

You're not Bullet Proof—I still remember my first call out. I had just topped out as a journeyman and we were getting a strong October wind storm. We had lights out everywhere. I was feeling great, bullet proof, until I was called to the substation and operated the wrong switch causing a very large 34.5 kV fire! In time, all of us young journeyman learn that we aren't bullet proof. In truth, however, understanding the fact that we are not bullet proof today, instead of after an incident, is an executive skill worth learning. While no one was hurt by my switching error, it caused some additional work for us that night and some additional 'trouble' for me. What are you doing to rid yourself of the bulletproof attitude?

At the time of this writing, our daughter is just a few weeks away from finishing the fourth grade. She has had a great year. And, with the help of her mom (and sometimes dad) and the larger team, she is doing very well in school. Yet with each new year, there are new executive skills to learn in order to continue success. What are your executive safety skills? And, what new ones will you need to develop for continued success?

30 Hazard Intelligence
Four Intelligences That Can Change Your Safety Culture

Have you heard of Chris Langan? Today he lives on a horse ranch in northern Missouri, but the path to this ranch was not easy. Chris was born in San Francisco in 1952. As an infant, his mother moved him to Montana. Chris never knew his birth father, and his mother remarried about the time Chris started grade school. The family lived in extreme poverty and Chris in abuse, from the stepfather. The abuse continued until Chris was 14. At that time, Chris took up weight lifting and threw the stepfather out of the home. Chris' high school years were spent in mostly 'independent study.' He did try his hand at college, first Reed College, then Montana State University, but didn't finish. As an adult Chris has worked mostly labor intensive jobs such as construction worker, cowboy, farmhand, and firefighter. He also worked as a bouncer in Long Island, New York. Chris' story, though somewhat heartbreaking, may not be all that different from thousands of other men, yet with Chris there is one difference. Chris Langan is the smartest man in America.

Compare Chris to another 'smart man,' Albert Einstein. While Einstein may be the most noted brainiack in the United States, his life may have not been any easier than Chris Langan's. Einstein was born in Germany in 1879. As a child, his family moved several times, both within Germany, and to Italy and Switzerland. Einstein moved through school, graduating from college as a math and physics teacher, but was unable to find a teaching job, so accepted a position in the Swiss Patent Office. He continued to work toward a doctoral degree then moved back to Germany to teach at the University of Berlin. He married in 1903, had three children then divorced. He remarried in 1920, but his wife died about a decade and a half later. He remained in Germany until 1933, but left his homeland due to political reasons, eventually moving to the United States.

Hazard Intelligence

This is where the comparison ends. Einstein, of course, has become a common name for 'smarts'. Einstein published more than 300 scientific papers and nearly 150 non-scientific papers. He earned a Nobel Prize in Physics. He was bestowed honorary doctorate degrees in science, medicine and philosophy. Einstein was asked to be the first president of Israel when the country was formed after World War II, but declined. Langan, well . . . Langan doesn't hold any degrees, has not earned a Nobel Prize, has not lectured throughout the world and has struggled to publish any of his scientific papers. If you are thinking that this isn't a fair comparison, you are right. Albert Einstein's IQ was 150 while Langan's IQ is 190—so Langan is 20 percent smarter than Einstein! But chances are you have never heard of Chris Langan—why?

Malcolm Gladwell in his outstanding book *Outliers: The Story of Success* suggests that to be successful on the level of an Albert Einstein you need two things. The first is you need to be 'smart enough.' After this initial cut of a high IQ, it takes something more, something termed *practical intelligence*. Gladwell quotes psychologist Robert Sternberg, "Practical intelligence includes things like knowing what to say to whom, knowing when to say it and knowing how to say it for maximum effect." He continues, "It's knowledge that helps you read situations correctly and get what you want." Some researchers have termed this 'social intelligence'. And the best thing to understand about this concept of social intelligence is that it is teachable!

So, what does this have to do with safety? Everything! Our organizations spend enormous resources working on our safety IQ, and they should. Safety IQ includes a keen understanding and knowledge of the safety rules, safe work practices, proper standards and operating procedures. And, organizations spend considerable time studying and thinking about culture, and the effects of culture on safety results. But, we generally don't dedicate time or resources to teaching and exploring Hazard intelligence—the ability to read a job site, hazard or changing situation correctly and perform our work injury and incident free.

In fact, four key hazard intelligences, which are teachable skills, might just add up to safety success and cultural change; let's take a look:

Safety Awareness—In 1995, I began an apprenticeship in distribution line work. The training program is intense and comprehensive. It was a good combination of hands-on learning with textbook modules. We had to test proficiency on a number of key tasks like chain saw use, grounding of overhead lines and termination of underground cables. What we didn't learn in the hundreds of modules was safety awareness. After all, is it okay to understand how to properly operate a chain saw, if I can't identify other site hazards that could cause a significant injury? Safety Awareness is one of the cornerstones of hazard intelligence. How well do you teach and evaluate this key skill?

Job Planning—A few years ago, as an area safety professional, I was called to a job site after an electrical contact. The worker who touched the 12,740 volt line was very lucky, he was at the hospital but would make a full recovery. In speaking to his other two crew members, I learned what had happened. The crew had energized a section of line, then took a break for lunch. During lunch, the lineworker who made contact took a phone call from his cell phone. Immediately after lunch, he went up in the bucket, completely forgetting the line was energized just thirty minutes earlier. The phone call was about his daughter, who was having a hard time. So, the injured worker's mind was not on his work. And, the crew completely failed to plan. For the most part, we teach our crews to review a job before it starts, and that's it. Jobs change and work progresses, job planning is a hazard intelligence that needs to be taught, job planning at the beginning of the job and throughout the day.

Peer-to-Peer Feedback—This year I coached my son's 9- and 10-year-old baseball team. It was a good group of players and parents, but one of the biggest challenges we had was getting the players to talk to each other. Everyone who has played baseball, softball, or any sport really, understands the importance of talking to each other—communicating. In baseball we call that 'chatter.' On our baseball team, you could hear crickets chirping, and many job sites and work floors are the exact same way.

In May 2010 *Bloomberg Businessweek* published an article called "The Peer Principle." This article reported research comparing organizations with good safety records to those with excellent safety records. One key finding read, "Peer accountability turned out to be the predictor of performance at every level and on every dimension of achievement. The differences between good companies and the best weren't that apparent when it came to bosses holding direct reports accountable. The differences become stark, however, when you examine how likely it is that a peer will deal with a concern." Peer-to-peer feedback, or 'chatter' as I like to say, is a significant key to safety success. It is also a hazard intelligence that must be taught.

Remaining Uncomfortable—Do you remember when you started driving? Or better yet, do you have a child who you are teaching to drive? We start out driving under the speed limit. Both hands are firmly pressed on the wheel. The radio is turned off, and the cell phone is safely tucked in the backpack in the backseat. We are uncomfortable and driving demands our maximum attention! Fast forward two years, the same driver has a big gulp soda in one hand and the cell phone in the other. They are driving faster than the speed limit with the radio blasting, their legs are controlling the steering wheel—they have no fear. One of the best practices to teach your organization is to stay just a little uncomfortable—staying well within safety rules and safe work practices clearly understanding the consequences of not doing so. There are a number of ways to do this, through active participation in incident analysis, near miss reports, job observations, employee sharing, etc. The key is that organizations have the hazard intelligence to remain comfortably uncomfortable—and safe.

In the end, IQ matters. Teach and train on safety rules and safe work practices—nothing can replace a high safety IQ. You, and your organization, must make this first cut. But once the first cut is made, we need to consider what makes safety even more effective; what can help take your organization to the next level of safety performance and possibly transform your safety culture. And, just as social intelligence is a clear indicator of life success, a keen sense of hazard intelligence may just be the ticket to safety success as well. Learn, and teach well!

Source:
Gladwell, Malcolm, *Outliers: The Story of Success,* Little, Brown and Company, 2008.

Grenny, Joseph, "The Peer Principle," *Bloomberg Businessweek—The Influential Leader,* May 2010.

31 Finding the Gorillas!
How Leaders Deal with Attention Blindness

I would imagine that May 4, 2012 started as a typical Friday for Jane. For a young teacher at Cedar Creek Elementary School in Kansas City, flipping the calendar over to May signaled that the school year was nearing the end. And Jane, along with her 13-month-old baby, would soon be able to enjoy the summer together. With only a few weeks left in the school year, things were getting busy and hectic. It was a beautiful day in early May and temperatures would be hotter than normal, reaching the low 80s. As usual, Jane dropped off her baby at day care then went to school. After another exhausting and chaotic day teaching elementary-aged children, Jane returned to her car. It was 4:30. Her life changed forever.

In her back seat, lifeless in the car seat where Jane had safely strapped the love of her life some nine hours earlier, was her baby. She reacted, yanking her child from the car. She accessed vital signs, nothing. Frantic, she called for help and begin to administer CPR. Nothing. Emergency services arrived and took over but it was too late. Jane's baby was dead.

A few months ago I was enjoying dinner with a client when they asked me a question, "Matt, what is the biggest safety issue facing utilities today." I casually cut my steak and stabbed the piece with my fork. Before putting it in my mouth, I said, "Failing to find the gorillas." Then I took the bite. They both looked at me like I was crazy!

There is a concept called *attention blindness* and it is best described through a video that you may have seen. Before you watch the video, viewers are asked to count the number of times that a certain team passed the basketball—generally it is framed as a competition between participants with the winners being the ones who are able to accurately count the number of passes. Once the video cues, there are two teams, a team in white jerseys and one in black jerseys, passing basketballs. Viewers count so intensely that a very high percent of participants miss the fact a gorilla walks onto the screen, pounds his chest then walks off!

So, how is a young mother, a teacher in fact, suddenly blinded so that she leaves the most important thing in her life, her baby, in the back seat to die? She does this in the same way our workers miss hazards that cause near misses, injury, and unfortunately, even death. We focus so intensely on one thing and miss the gorilla pounding his chest in the middle of the screen—attention blindness.

Attention blindness is literally a life and death phenomenon. And, how leaders recognize and communicate attention blindness might just be the most important issue a leader can address.

So here are four principles to teach that raise awareness around attention blindness and four ways to help prevent it from happening.

In order to raise awareness around attention blindness, consider the following:

Attention Blindness Is Real—First, understand that attention blindness is a real and important concept. People can focus so intensely on one task (counting balls) that they miss other key events (a gorilla). To reinforce this fact, start by showing the video to your workers. That way they can 'experience' attention blindness first hand. (You can easily find it on YouTube by searching keyword, 'attention blindness.') Make sure you set up the video with a contest asking people to compete by counting the 'exact' number of passes by a certain team.

Attention Blindness Blinds Completely—Next, our supervisors and workers alike need to understand that not only can we focus so intensely on one thing that we miss other events; we are completely blind to these other events. Being completely blind means that we don't see what is right in front of our face, and in a hazardous job, missing this hazard can mean injury or death.

Attention Blindness Blinds All Things, Including our Most Treasured—
We are blind to the most important things in our life. I don't know the young

teacher at Cedar Creek Elementary School. My summary of her day was just my reconstruction of the day based on my experience as a dad, rushing kids to school with a dozen other things on my mind. Being married to an elementary school teacher for nearly two decades and having analyzed a number of serious injuries and events, including fatalities. But in truth, it doesn't matter. The facts around her day don't matter. What does matter is the fact that she was completely blinded by attention blindness. And, it took the life of her child.

It Can Happen to You!—I spent a number of years as a safety professional on a utility safety staff. I have reviewed injury and incident reports in front of hundreds of utility workers only to have those same workers shake their heads and say, "I'd never do that." But, just as it can happen to a teacher in Kansas City, it can happen to us. Knowing that attention blindness is real, it blinds completely, it can take important things from us and that it can happen to us at any time is half the battle in making sure it never happens.

After your workers understand and acknowledge attention blindness, consider these prevention techniques.

See the Entire Picture Before You Start—In many industries, including utilities, work rules and OSHA require that before you begin work you review the job, all work rules, PPE, energy sources and potential hidden hazards and make a plan. It is in analyzing hazards, making a work plan and communicating this plan to crew members or people on site that you should smoke out any gorillas. Make sure that you and your workers are planning and consider, if you have not done so already, to formalize a procedure around this planning process in order to make it even more difficult to miss gorillas.

Have Every Worker Watching Something Different—I started with the utility as a meter reader, then a lineman, and from there I was promoted to the utility's safety staff. On that staff, the most tragic incident that I had to review was a double electric contact that resulted in a fatality. The crew that day was made up of six journey line workers. Between all men on the job, there were decades of experience. The job was easy, to change out a three phase in line and lay the phases out on insulated hot arms. The only hazard on the job that had the potential to end a life was the 12 kV line. And, after less than a half hour into the job, the boom contacted the line when two workers were in contact with the truck. One died.

Upon review, all six were working to move the new pole into a position in order to frame it. When all of them concentrated on counting the same ball (moving and framing the pole) the gorilla appeared. When possible, assign each crew member the task of seeing or watching something a little different. Make sure you have identified all of the gorillas that can lead to immediate injury or death and make sure they are watched. Each job, each time.

Use System Two—Quickly respond to this question, "A bat and ball cost $1.10 and the bat costs a dollar more than the ball. What does the bat cost?" What more than 80 percent of college students answer is that the bat costs a buck. If you do the math, the bat is a nickel and the ball is $1.05!

In his book *Thinking, Fast and Slow*, Daniel Kahneman says this quick response comes from what he terms system one. System one is our effortless response due to habit. If you are like me, and eight out of ten college students, you used system one to quickly decide that the ball was a dollar. System two, on the other hand, is the thinking system, the problem solver, or the effort system. System one misses gorillas while system two finds them. Each day our workers go out and work in system one mode. Find ways to push them to think, so they can think, and in the process get all hazards identified.

Find Practical Solutions—Gorilla (hazard) awareness is the right start, but the next step is to employ practical solutions to find gorillas on each job. Here are some solutions to try. First, think of employing a safety stop. A safety stop is when you stop about every 60 minutes to quickly review the job, reidentify the gorillas, make sure they are accounted for, and then continue work. A safety stop is no more than a 90 second exercise, but it can save lives. Next, have crew members and supervisors perform job audits looking for gorillas and giving immediate feedback to crews. The final suggestion is for safety huddles. Each morning, take less than five minutes to huddle with your work group or crew and discuss likely gorillas. In most industries, changes due to weather, the job at hand, manufacturing schedules, vacation and sick leave of personal, season of the year, etc. At the end of the day, huddle again to quickly share any near miss or hidden hazard.

The day after the tragedy at the school, the school released the following statement, "Sharing this heartbreaking news and leading our school family toward a healthy acceptance of this tragic event are tasks that lay ahead of us." The hazards associated with today's world, both at home and at work, are many and the consequences for not properly finding and eliminating and controlling those hazards are unforgiving. Find the gorillas today—and be safer for it!

Source:
Kahneman, Daniel, *Thinking Fast and Slow*, Farrar, Straus, and Giroux, 2011.

32 Do You Have Safety With-It-ness?

In the December 15, 2008 *New Yorker*, Malcolm Gladwell published an interesting article on education. The article focused on how teachers are evaluated. The article is entitled, "Most Likely to Succeed, How Do We Hire When We Can't Tell Who's Right for the Job?" introduces a concept called with-it-ness; and asserts that today's most gifted teachers have it!

Gladwell writes,

> "Another educational researcher, Jacob Kounin, once did an analysis of 'desist' events, in which a teacher has to stop some kind of misbehavior. In one instance, "Mary leans toward the table to her right and whispers to Jane. Both she and Jane giggle. The teacher says, 'Mary and Jane, stop that!'" That's a desist event. But how a teacher desists—her tone of voice, her attitudes, her choice of words—appears to make no difference at all in maintaining an orderly classroom. How can that be? Kounin went back over the videotape and noticed that forty-five seconds before Mary whispered to Jane, Lucy and John had started whispering. Then Robert had noticed and joined in, making Jane giggle, whereupon Jane said something to John. Then Mary whispered to Jane. It was a contagious chain of misbehavior, and what really was significant was not how a teacher stopped the deviancy at the end of the chain, but whether she was able to stop the chain before it started. Kounin called that ability 'with-it-ness.'"

Gladwell, and researcher Jacob Kounin, simply define 'with-it-ness' as the ability to be proactive—to stop an event before it happens. And, they assert that with-it-ness, along with feedback, is the key trait to teacher effectiveness. Safety with-it-ness, or our ability to be proactive or to stop an event before it happens, is also the measure of safety effectiveness. So, the simple question is, do you have safety with-it-ness? Here are three simple questions you should ask every day in order to have a high level of safety with-it-ness.

First, what on this job can change my life forever? Having worked in the electric industry for nearly 20 years, I have had the misfortune to analyze dozens of electrical contacts and other serious incidents. What is amazing about many of these tragic events is that on most of these jobs, the energy source was the only hazard on the job that would change the workers life forever. And, it was the energy source that went unguarded or that the worker didn't take the appropriate action to protect himself against. Before starting any job, ask a simple question; what on this job can change my life? You may find an energy source, trench, fall exposure, vehicle traffic, etc. Usually however, there are only one to three things on each job that are 'major' life changing hazards. Find them. Take proactive steps to control them—work with safety with-it-ness.

Am I in a transition? Several years ago the quest to summit Mount Everest hit 1,000 official requests. When reviewing those attempts, statistics show 200 of the 1,000 climbers perished. The interesting thing for me is that of those 200 who died, 150, or 75 percent, died climbing down the mountain. One could argue that the focus and energy and planning was dedicated to reaching the summit and climbers lost sight in transition, underestimating the focus and planning that phase of the climb demands.

Our jobs are often similar with transition situations often leading to incidents and injuries. Our equivalent of descending the mountain might be when a construction job is finished and the only task is cleaning up, or backfilling or climbing down off of a pole or elevation. Transition is simply the last task before break or lunch. It might be the day before a long vacation. Having safety with-it-ness means we actively look for transition situations knowing that there is added danger in these times. If you don't believe me, think about those 150 climbers. Jeff Evans, a man that has successfully climbed and descended Mount Everest says it this way; "Reaching the summit is optional . . . going home isn't."

What's new? The Occupational Safety and Health Administration published the following on an OSHA facts sheet, "Young workers, ages 14–24, are at risk of workplace injury because of their inexperience at work and their physical, cognitive, and emotional developmental characteristics. They often hesitate to ask questions and may fail to recognize workplace dangers. OSHA has made young workers a priority within the agency and is committed to identifying ways to improve young worker safety and health." In other words, they are new, and because they are new, there is added risk of incident and injury.

But being new doesn't just apply to young workers. Having an added risk of injury or incident applies to new tools, new equipment, a more seasoned worker in a new job, etc. Safety with-it-ness means you and your organization understand the added risk of 'new' and take pro-active steps to review and train. This small amount of time making 'new' more comfortable and getting your workers more comfortable with 'new' tools and equipment can save more time later, and can save lives.

In short, good teachers are able to stop events before they occur; and for us to make it through each day safe, we need to employ that same skill set—or as I like to say, we need safety with-it-ness!

33 Chicken Rings
What's in Your Circle of Safety?

Have you ever heard of a chicken ring? Even though I have earned a journey lineman card in distribution line work and like to run excavators, chainsaws and skid loaders, I'm still a city slicker. So, when I was given my first chicken ring, I didn't have any idea what it was. It turns out chicken rings have at least two primary purposes. These small plastic rings, which are about the circumference of a quarter, are used for chickens. The rings can be quickly slipped on and off of a chicken leg or neck to mark it. Apparently if you are a chicken and earn a chicken ring, it's a good thing—you get to live another day. The second use of a chicken ring, interestingly enough, is for safety.

My first chicken ring was given to me by Tom. Tom was a legendary electric foreman. His demands were high, he motivated his men and he had a knack for safety awareness. He purchased green chicken rings by the bushel. As Tom would conduct a job briefing or safety talk, he would insist that his men take a chicken ring or two. These rings would we slipped on hard hats, boots, door handles, steering wheels, bucket controls, etc. The point of the ring was to raise safety awareness—awareness to the Circle of Safety.

This simple green chicken ring is a great reminder of a simple yet powerful habit, a circle of safety. In the utility business and any safety sensitive environment, there are several times that we are taught about the 'circle of safety.' In our defensive driving classes, we are taught that our large trucks have a circle of safety. It's the zone around our vehicles where smaller cars, or even people, can sneak into a blind spot. We are reminded to constantly monitor these blind spots to prevent an incident. After a job is finished, yet before one hops in the truck to drive to the next job, we are asked to perform a circle of safety. This is a complete walk around of the work site and vehicles to make sure our load is secure, bin doors closed, and all tools are picked up and properly stored. We need to perform a circle of safety before we

Chicken Rings

back our vehicles, since we can't see what's behind us. The point is simple; a circle of safety is an effective safety tool. A high level of safety awareness is needed to keep one focused on practices like the circle of safety, and a chicken ring is an effective reminder. To that end, let's take a quick look at the five key things to think about regarding chicken rings and circles of safety.

Five Keys to an Effective Circle of Safety:

1. Remember Changing Conditions—Joe was in a hurry and he needed to pick up some materials for the next job. He buzzed into the works headquarters, parked his bucket truck and dashed inside. He quickly returned with a handful of material, which he put away in the driver's side bin. He thought about a circle of safety but reasoned that he had been there less than five minutes, so he climbed into the cab and backed out—backing directly into another truck that had parked behind him, right in his blind spot. This story is completely true and hundreds of incidents like this happen every day in our industry. Remember, conditions change by the second and a simple walk around, or circle of safety, is the best way to prevent these events.

2. Look for and Eliminate Hazards—Everyone who has worked for any time in a hazardous field has a tragic story. As a former safety professional for a utility, I have a number of stories, but the one that stands out is when a crew inadvertently raised a boom into a 12 kV phase. When the phase made contact, a worker on the ground was getting material off the truck and he received a fatal electric shock. Between the linemen on that job, there was probably over 100 years of experience. Everyone on the job saw the overhead wires. Everyone saw where the truck was parked. The only thing on the job that could have caused serious injury or death was the energy source—it was recognized but not eliminated with rubber cover. The circle of safety encourages that we not only identify hazards, which is the first key step to safety, but it requires the next step, to eliminate hazards. Through this circle of safety, maybe the next generation of utility workers won't have tragic stories to share like our generation does.

3. It's Not Just for Trucks and Poles—For years the utility industry has promoted a circle of safety when backing a utility truck and before climbing a pole, but this concept isn't just for trucks and poles. In fact, the principles of a circle of safety are easy; before you begin a task, walk through the work area, identify and eliminate hazards, then proceed. If we can use this easy, 'before you begin a task, walk through the work area, identify and eliminate hazards,' formula on each job and each task, we have a formula for safety success and injury prevention that can't be beat!

4. Expect the Unexpected—A number of years ago, an old troubleman completed a job in the backyard of a home. He then walked from the work location to the front yard and to the street, where his truck was parked. He knew he would need the same tools for the next job, so he set his tool in the passenger floor board. He did not walk around the truck, instead, jumped in and drove to the next job. As he got out of his truck, he heard crying—it was coming from the back of his truck! He ran to the rear of the truck to find a young boy lying there in tears.

It turns out that while the troubleman was in the backyard working, the young boy was playing on the truck. When the troubleman returned, the boy panicked and froze, remaining on the truck. The incident could have been tragic if the child had chosen to jump off the moving truck. And, the whole incident could have been avoided by a simple walk around.

5. Chicken Ring It—One of the keys to performing a circle of safety is awareness. Utility workers need constant awareness, and one of the best ways to create awareness is through a visual reminder—this brings us back to the chicken rings. After 40 years in the utility business, Tom is retiring this year. It is a running joke that his biggest accomplishment over these four decades isn't in the utility world, instead, he single-handedly kept the chicken ring company in business by purchasing thousands of these rings. I'm not sure if that is true, what I do know is that he had a great impact in the safety of his workers by teaching the circle of safety and handing out chicken rings as a visual reminder. It's not about Tom, or the chicken ring. It's about keeping our people safe through simple, yet memorable awareness activities, so that each of our people can leave work as healthy as they arrived. What is your chicken ring?

34 Check Down

How Football Quarterback Can Make Us Better at Safety

A college quarterback moves under center getting ready for the snap of the ball. It's a Saturday in the fall and it's game day. The crowd is going crazy—it is so loud 'one can't hear themselves think.' The quarterback barks out some commands. Then, he signals for his tight end to go in motion from the left side of the formation to the right. Or, instead of the tight end, this time it is the wide receiver, running from his original spot on the right of the formation across to the other side. Have you ever wondered why this happens? Why does a quarterback put players in motion? In short, so that he can read the defense.

By placing a wide receiver, tight end, or running back in motion, the defense will often have to 'show their hand.' How the defense reacts to an offensive player in motion will often give the quarterback insight on the defensive scheme. The quarterback is also looking for clues to answer questions like, is the defense matched up man-to-man on the outside receivers? Are defensive players lining up to blitz? Is the defense in a nickel package or dime package? Are there extra defensive lineman to give run support, or more cornerbacks to cover a passing situation? This information tells the quarterback if the play that is called has a high likelihood of success or not. But, more times than not, right before the snap of the ball the quarterback will do something unusual. He will step back, he will check down.

Generally, often right before the snap, after reading the defense, and evaluating the situation, the quarterback will move away from the center. He may have ten or twelve seconds to snap the ball, and he'll use them wisely. He can look over to his bench, where coaches have headsets getting advice from other coaches perched in press boxes high above the stadium. The coaches on the sidelines are talking to the quarterback through hand signals. The quarterback will often audible to a

What Utility Safety Leaders Do

modified play, reset his team, then move ahead with the play. In the course of about 25 seconds, the quarterback will receive a play from the sidelines, communicate that to his team, then set his team. He will read the defense but before the snap, he'll likely check down. In so doing, compare the defense to the play called. Survey the defense by putting a player in motion. Get advice from his bench. Read the defense on more time, then move ahead. And, what might you ask does this have to do with safety? Actually, quit a bit!

The principles of a check down and the core principles leading up to that snap, every snap, seem to align nicely with what we and our teams should do before we 'take a snap' (in our business, that means 'go to work'). In fact, there are ten check down principles that we should make sure our team employees, before we take our snap.

The 10 Check Down for Safety Principles That You and Your Team Should Do Before You go to Work:

1. Stay in Game Shape—Before a snap is taken, or even considered, players must first think about their bodies, and how good of shape they are in. In fact, all players are athletes. And, from the kicker, to quarterback, to nose guard, these athletes buy in to a holistic program involving exercise, weight training, and diet. That said, our workers are athletes too. After all, they must use their bodies for 40 hours a week, 52 weeks a year for over thirty years. To do that you had better be an athlete in great shape. Over the last few years, ergonomics, pre-work stretch programs, proper lifting, and mechanical aids have been introduced to help our athletes stay in game shape. What is your 'game shape' program?

2. Study Film—Before each college football game, each player spends hours watching and studying game film. The purpose is to understand what the other team does, how they do it, and what can be done to counteract that team. For our work, we probably do not spend enough time 'watching film.' With today's modern cameras, smart phones, and access to computers, we should ask our supervisors and workers to take pictures or video. Just as college football teams study film, we should employ aggressive practices of capturing 'film' and pictures. We should capture pictures of hazards. Grab video of near misses and incidents. We should ask supervisors to get video of jobs done well and safe work rules followed. Actually, it wouldn't be unreasonable to have five minutes of film study for our work teams each day. Think about how that could positively impact your team—and raise awareness to a new level.

3. Huddle (Have a Plan)—Before each play, a football gets in a circle and communicates a carefully crafted and practiced play. Before we go to work we should do the same. Just like in football, we need to circle review the job. Each day we need to cover the work at hand, PPE require, safety work rules that apply, any energy sources and means to properly control them, along with any special precautions associated with the job. A football team won't snap the ball without a huddle. Why would our team start work without one?

4. Know Your Position—When the center snaps the ball to the quarterback, every player on the field has both a position and a specific task. Success of any particular play depends on how well each player executes that specific role. Our job sites and work tasks should be no different. Each person needs to know and understand their role, responsibilities, and receive the appropriate training for that position. Football legend Knute Rockne said, *"Football is a game played with arms, legs and shoulders but mostly from the neck up."* And, so is safety.

5. Survey the Situation—At the heart of a check down is the quarterback's ability to survey the situation, understand the information he is receiving, and adjust for his team's success. Our work should be no different. Before beginning work, survey your situation (job site, surroundings). Walk through the job site or work environment assessing hazards and unusual elements that need to be address. Make sure that you have the right play called.

6. Take the Time It Takes—Immediately after a play, the football official will set the ball and start the play clock. At that time, a 25 second clock begins, which signals that the football team has 25 seconds to snap the ball. In our work, we don't have a play clock. Instead, we have the opportunity to take whatever time we need to make sure we are ready before we snap the ball (begin our work). Take the time that is needed to run the right play—there is no play clock in our work.

7. Use Timeouts Wisely—From time to time, after the quarterback 'checks down' the defense will throw him a curve ball—a formation that he has not seen or one where he feels he can't make the right adjustments. In those rare situations, he can call a time out. As you know in football, a team is only allowed three timeouts per half. In our work however, using timeouts wisely means we call frequent 'timeouts.' We sometimes call these frequent time outs safety stops. These are times when something doesn't look right, the job has changed, a new crew member has arrived, we finish one part of the job and starting on a new piece, or we simply want to take 60 seconds to reset before we move forward. In our work, call frequent timeouts!

8. Regroup at Halftime—Football games can be won or lost at halftime. During this time adjustments are made in an attempt to perform even more effectively and to get better results. In our work, we have frequent 'halftimes.' We have lunch, breaks, moving from one job to another. After each stoppage in work, make sure that we plan our work accordingly and check down to ensure our safety.

9. Learn from Last Week's Game—After each game, football teams review the game film. In fact, in most college programs, each player is graded on each play with each player receiving a game grade. To that end, spend a few minutes after each job, or at the end of each day, talking about your work. What went well? What new hazards were present? What has changed? What can we expect tomorrow? A football team misses an opportunity to learn if they don't review the most recent game, and we do the same if we fail to review our day.

10. Play to Win!—"Some people think football is a matter of life and death," said Bill Shankly, "I assure you, it's much more serious than that." Safety in our work is the same—literally a matter of life or death. One way to better prevent incidents and injuries is to practice a check down, and use each of these principles to raise worker awareness.

Walter Payton once said, "I want to be remembered as the guy who gave his all whenever he was on the field." Those that remember Walter Payton have fond memories of his grace, athleticism, and work ethic. Do you do the same when it comes to your safety and that of your team? Check down today—and every day. Once you do, you can hike the ball and go to work!

35 Unreasonable Leadership

Challenging What We Think is True about Safety Performance

Would it be possible to set and achieve the goal of weighing what you did when you were 14 years old? That would be not only unreasonable but ridicules, yes?

In Katie Couric's insightful book *The Best Advice I Ever Got*, speed skater and eight-time Olympic medalist Apolo Ohno shares this:

"The 2002 and 2006 Olympic Games both had many instances which I was the strongest, the fastest the most fit and also the skater with the best strategy but something happened and I didn't come in first. Either I slipped or someone bumped me. Something seemingly out of my control happened and I didn't win the race. I complained to myself, 'man that kinda sucked' because I felt like I was the best person but I didn't win.

But in the end it wasn't really about the win or the loss. In 2002 I weighed about 165 pounds and I leg pressed approximately 1,400 lbs. In 2006 I weighed about 157 pounds and leg pressed about 1,500 lbs. My strength-to-weight ratio was a lot higher (in 2006). Four years later in 2010 I vowed to weigh less than 150 pounds. I wanted to race at 147 pounds so I had to totally change my mentality about what was possible from a physical perspective. To put that in perspective, I hadn't weighed less than 150 since I was 14 years old. And here I was going to be almost 28. I ended up racing at 141 pounds and leg pressing almost 2,000 lbs, so that to me was a testament to strength of my mind and will.

You can accomplish whatever you set out to do even when people think it can't be done."

I think Matt Goldman, cofounder of The Blue Man Group, is saying the same thing as Apolo Ohno only that he says it differently. In 1987, as Goldman was finishing college at Clark University in Worcester, Massachusetts, he and two friends came up with the concept of the Blue Man Group. For some observers and critics this was a 'strange' way to put on a show—and one that would clearly not last. Think about it, the show featured grown men in a blue suits who don't talk to the audience but attempt to relate in other ways. Goldman and his partners received all sorts of feedback too. Some people told them that the show would never work, some said it was too funny to be a hit, some said it was not funny enough, and the list went on and on.

But Goldman and the Blue Men didn't listen, instead they honed their act, and people noticed. Now, some two and a half decades later, the Blue Man Group is one of the most popular and noted acts anywhere. They have played shows across the world, played on stage with famed performers, appeared on numerous TV shows, and met countless dignitaries and world leaders including the Dali Lama. Goldman says, "We heard all of the reasons it was not going to work, but guess what, it did work. I wanted to be crazy and I advise you to be crazy, to be weird... to be unreasonable. That's my favorite one. People are always saying 'Oh come on, be reasonable' and I want to shout 'No, I don't want to be reasonable.' I want to be completely unreasonable. I want to change the world..."

The Unreasonable Path Forward—"All truth passes through three stages," wrote Arthur Schopenhauer. "First, it is ridiculed. Second, it is violently opposed. Third, it is accepted as being self-evident." The role of the leader is to challenge their respective organizations to grow and meet goals—sometimes seemingly ridiculous goals. One great tool to such growth and innovation is to be unreasonable. Like any tool, it needs to be used wisely. Below are some tips, or a path forward, on harnessing the power of unreasonableness.

Define the Unreasonable Playing Field—Safety has some notions, truths, or principles that are largely accepted as fact. Some of these 'truths' are: It takes ten years to really change a safety culture—why can't we do it in 365 days? Zero incidents is not achievable—why can't zero be both achievable and sustainable? Training doesn't change behavior—why shouldn't one training class change behavior? Finally, since a study (entitled "The Peer Principle" by *Bloomberg Businessweek*, May 2010) found, "In the area of safety, our study found that 93 percent of employees say they see urgent risks to life and limb and yet less than one-fourth of those who see concerns speak up about them. Rather, they wait for bosses or others to take action," why can't peers hold peers accountable?

Being unreasonable is simply taking a notion that is largely accepted as fact, and turning it upside down (setting the course to 'self-evident'). A notion that says one can't speed skate at their pre-driving weight. Or, a concept that says men in blue suits who don't talk will not be a hit. Or the perceived truth that people can and will get hurt at work. A leader's job is to capture or write down as many of these notions as possible then decide which one or ones should be challenged.

First, Believe—Henry Ford coined this phrase, "Whether you believe you can or believe you can't, you are generally correct." Apolo Ohno believed, despite how unreasonable it might have sounded, that he could race at 147 pounds or less. Goldman and his partners believed they could make it, despite the feedback from some industry leaders. As a leader, you first must believe that what you are saying is true. For example, if you set the goal to change your safety culture in 365 days, ignoring common counsel that it takes much longer, you must in your core believe it. By the way, if you believe you can change your safety culture in 365 days, you are generally correct!

Get Counsel from a Council—When Goldman and the Blue Men were just starting out they formed an informal council of close friends and trusted industry experts. They wanted and needed people whom they trusted and whom they could talk to about their show. They wanted a 'safe group' to talk candidly about the show, what works, what didn't work, and how to improve it. They didn't listen to their critics, instead enlisted help from those they trusted.

When you choose the unreasonable path, you will have critics so before taking this first step establish a small team. This advisory group can make your ideas better, give you key insights on your goal and provide counsel against critics.

One Step at a Time—There has maybe not been a more unreasonable goal than when now former General Electric CEO Jack Welch announced the 'number one or number two' goal. It meant that every GE business line had to be first or second in their respective business, or they would be sold! I can still imagine the sinking feeling in every manager's stomach when this was announced. But, as Welch led the company, this was the only unreasonable goal. It may have been the only goal and everything else was measured against it.

I like the old management saying, "When you're the hammer, everything looks like a nail." We have all known, and maybe even worked for, the 'boss' who solved every problem the same way—and maybe the solution was always the hammer. The point is that using only one tool in the toolbox is ineffective. Leading with unreasonableness is also a tool. Consider one unreasonable goal at a time, like Welch. Once that goal is achieved, then move on to another.

Be Mindful of Harmful Unattended Consequences—Rob Norton, former economics editor for *Fortune* magazine says that unintended consequences is, "the actions of people always have effects that are unanticipated or unintended." Setting an unreasonable goal will also have unintended consequences. Some of these will be positive, like innovation and creatively. There will undoubtedly be some negative consequences too. In the area of safety, explore whether there are harmful unintended consequences. For example, if you set the goal of zero incidents, injuries may go unreported in order to make the end of the month report look good. That is not your goal, hiding injuries, your goal is zero injuries.

Plan the Plan—At the end of the day, your unreasonable goal is a corporate or business line strategy. Just like those endeavors, it needs to be planned, coordinated, communicated, measured and aggressively pushed. So, plan your unreasonable plan.

In Closing—General Eric Shinseki has said, "If you don't like change, you're going to like irrelevance even less." A few years ago researchers put a Pike fish, which is a fish that aggressively eats other smaller fish, in an aquarium with other smaller fish.

The only 'catch' was that researchers inserted a piece of glass between the smaller fish and the Pike fish. Immediately the Pike fish dashed toward the smaller fish, only to be repelled by the glass. Again and again the Pike fish smashed his nose into the glass trying to reach his dinner. Eventually, the Pike fish gave up, sinking slowly to the bottom of the aquarium. At this point researchers removed the glass barrier. Smaller fish began to swim throughout the entire space, even at times, brushing the Pike fish. A few days later, the Pike fish died of starvation.

Today organizations who hold on to 'Pike Fish' beliefs are in trouble. We need leaders who can be unreasonable, pushing their companies and the field of safety into new results. Doesn't that sound reasonable?

Source:
Couric, Katie, *The Best Advice I Ever Got: Lessons from Extraordinary Lives*, Random House, 2011.

Grenny, Joseph, "The Peer Principle," *Bloomberg Businessweek—The Influential Leader,* May 2010.

36 Safety's Innovation Cycle

What Safety Leaders Understand about Innovation

How Innovative are You?—Grab a piece of paper and a pen and take this quick quiz. Set a timer for two minutes and write down as many uses of a paper clip as you can. How many did you get?

If you decided on between one and ten uses of a paper clip in two minutes, your mind and creativity has been stifled by 'corporate think', which has slowly drained your innovative thoughts over time. If you thought of between 11 and 19 potential uses for a paper clip, you have innovation potential. If you thought of more than 20 uses of a paper clip you are either an Eagle Scout, or an innovator, or both!

Why Is Innovation Important to Safety?—Here are a few facts, "American corporations spend more than $50 billion annually on training. Yet, there were 4,609 fatalities in the workplace in 2011." And, the five-year trend on fatality rates is more or less flat over the last five years. In other words, minus a 'disruptive' innovation in your safety program, we will more or less get what we have gotten over the last five years. So, how good you are at thinking of uses for a paper clip, innovations in safety, might just determine your future trend line for injuries.

Forecasting Safety Innovation—What does J. C. Penney and Sears have to do with Innovation? To begin 2013, business analysts listed five companies that are on the watch list to 'go out of business' before the end of the year. Among these five were J. C. Penney and Sears. Both of these companies have a rich American history.

Safety's Innovation Cycle

Sears, officially known as Sears, Roebuck & Co., was founded by Richard Warren Sears and Alvah Curtis Roebuck in 1893. Sears started as a mail order catalog, and in the mid 1920s began opening 'brick and mortar' stores and today operates over 2,000 stores across the United States. James Cash Penney first opened The Golden Rule store in April 1902 in Kemmerer, Wyoming. The concept grew to 34 stores throughout the Rocky Mountain region a decade later. Today, J. C. Penney has more than a 1,000 retail outlets. Both retailers have struggled through the last decade, including Sears being purchased by Kmart in 2005. The reason for this struggle, in part, failure to realize innovation in the retail markets.

Four decades after James Penney opened his first store, a 22-year-old man named Sam Walton received his first job in retailing, at J. C. Penney in Des Moines, Iowa. He worked there for 18 months, learning the retailing trade. From there he moved to Arkansas to open a Ben Franklin store. In 1962 Walton opened his first Wal-Mart with an innovative concept, discounting products so that profits are less per product, but increased sales makes up the difference.

Today, Wal-Mart has nearly 9,000 stores and is the major competitor to Penney's and Sears. Both Penney's and Sears were well positioned to adopt Walton's innovative model some three decades ago and actually stifle the growth of Wal-Mart, but neither realized that this new model was their biggest competitor—neither store adopted this innovative model until it was much too late.

Safety's story is different than Sears, Penney's, and Wal-Mart, yet there are lessons to learn. In the 1970s OSHA's enactment was a disruptive innovation for safety. This placed structure, oversight and accountability across all industries. In that same decade (1979), E. Scott Geller coined the phrase Behavior Based Safety (BBS). Geller, along with other safety leaders, infused innovated concepts that are still the fabric of our safety programs today, including safety committees, employee observations, and management accountability. Yet, both of the major innovations that govern safety today, OSHA's structure and BBS, are over thirty years old and fatality trends today are flat. In short, what got us to this point in time will not get us to that next level of safety excellence. Safety is ready for that next innovation, and you and I will be the ones to bring to a reality.

Here are Seven Steps to Begin Innovating Safety.

1. Sponsor—Maxwell Wessle, in his February 2, 2013 *Harvard Business Review* blog post "How to Innovate with an Executive Sponsor," writes some insightful words:

> *"Meaningful innovation requires sponsorship. It always has. In 1959, one of the most important economists you've never heard of—Edith Penrose—pointed out as much by chronicling the nature of firm evolution. Penrose explained that all things equal, a firm's history determines its future. We seed our organizations with resources—people, capital, and equipment—and those resources have productive value in certain areas. Maximizing their value will naturally lead us to make the next decision and the next decision and so on. At its core, Penrose's idea is the reason innovation requires sponsorship. Without the foresight and intervention of senior leadership, the firm will simply concentrate on the opportunities that it was destined to concentrate on. Middle managers with limited resources and set evaluation metrics will simply operate in a predictable fashion.*
>
> *The difficult truth is that sponsorship as it's traditionally considered inside of large organizations is a double-edged sword. Sponsorship overcomes organizational roadblocks but often comes with a set of inherent limitations. Senior executives focus on big issues every day, when they turn to innovation they need their novel solutions to be equally as large. That's because nominally, the execs that matter inside of large organizations are used to moving the needle. So when it comes to innovation, executives are trained to value acquisitions, high profile product launches, and anything else they might use to surprise their analysts; without such surprises they can't generate unforeseen growth and placate investors."*

To break this 'predictable fashion' one or more senior leaders should sponsor innovation. To do this, they need to consider the following steps:

▸ Name the safety innovation sponsor(s).

- Set the vision—many times this is called the 'specific end'; where the organization needs to go, or the direction of the innovation.
- Lead with patience—many of these innovations won't immediately 'move the needle' but with the proper leadership, can be meaningful.

2. Create—How many uses of the paper clip you and your fellow coworkers identified show how easy or hard this step will be. In short, however, the first step is actually identifying a process to create. This process should include key items like when, who (the 'who' should include key groups like managers, first line supervisors and safety committees), where, and how often.

Creating can take the form of old fashion 'brainstorming' but each month creative time can be spawned by key questions. In his article, "8 Ways to Be Innovative (Even if You're Not)," Jeff Haden, *Inc Magazine*, lists a number of key questions that can prompt your group. These include; Imagine the worst that could happen, Play the "Why?" game, Pretend you just ran out of money, Pretend there are no rules, Pretend you only have five minutes to solve a problem and Imagine perfection. He goes onto say that one should 'Take a field trip . . . and borrow away'—so set up a means to identify and use best practices.

3. Evaluate—If creating is hard (think of the paper clip exercise) evaluating ideas is even harder. What makes a good idea or concept to one person may not even make sense to someone else. And, how do you align all ideas on an equal playing field so that they can be evaluated equally. Generally, keeping the idea with the creator is important. So, the best way to foster innovation through the creation phase into the evaluation phase is to set a structure for each group to evaluate ideas and decide which ones make the cut for what comes next, mini testing. To 'make the cut' in the evaluation phase think about the following:

Publish a formal innovation evaluation matrix (IEM) and have this form filled out for each idea that you consider worthy of next steps. The matrix can include questions like:

- How does the idea help solve the biggest problem you are having?

- How would this concept be implemented?
- Can you obtain the resources (skills, budget, time, etc.) to implement?
- How will results be measured?

4. Mini Test—This is where the excitement begins. Taking an idea through an evaluation process, then into the 'field' is, after all, the goal. If the IEM was comprehensive, it will provide the blue print for your mini-test. The information from the IEM can be added to the mini test guide, or MTG. This is a more comprehensive review of the testing parameters and include the traditional, who, where, when, how many, how long and data that will be collected along the way.

5. Measure—Data, data, data! Your sponsor will lead with patience, but at the end of the day, data drives innovation and decisions to scale concepts and ideas. To that end, deciding how to observe and collect data both before and after the innovation is tested is important. This holds especially true if you will be asking for additional budget dollars, or other resources to support the innovation long term. Just as we have an IEM and MTG, setting up a more formal measurement and data structure is a good idea.

6. Share—when I worked as a safety professional for a utility company, we'd bring together all of the local safety committees, over 20 of them, each summer to share ideas. This was a great way to trade innovations and give energy for the remainder of the year. Sharing innovation can be just that simple, a time when each group who is in the systematic create mode can come together to trade ideas and share accomplishments. That said, I also think that organizations should establish a more formal means to share. Asking each group to systematically share mini-tests and their associated evaluations with a small innovation committee, will allow ideas to be centralized, allow one team to review and combine ideas so you are not doubling resources and combine ideas to get even better results.

7. Scale—Scaling is taking the innovation company wide, and beyond. The secret to scaling is this equation; Q (quality of proposal) x A (acceptance of the proposal) = R (results). To this point, the innovation has been about the 'Q', the quality of the proposal, and the associated results. When the innovation moves to this stage,

Safety's Innovation Cycle

acceptance is the most important part of the equation. Write an acceptance plan to ensure the innovation equals results.

In closing, think about this. Some years ago, a young computer innovator spoke to a group of like-minded innovators. From the inside of his jacket pocket he pulled a folded piece of notebook paper. As he unfolded it, he explained to the audience that he and his company were pushing, and that someday a computer would be exactly like this paper. He unfolded it and explained that one would simply need to touch the screen to pull up email, surf the web, type a document, or whatever. The speaker was Bill Gates and today technology is close to his vision. Smart phones and tablets are pushing limits and screaming toward that vision. What is the next big safety innovation? With fatality trend lines flat, your company (and the safety community) needs it now, and this model for safety innovation may help pave the way.

37 The Secrets to $1,000 an Hour Work

What Leaders Know about Spending Time and the Value It Brings

Have you noticed the huge disparity in some hourly rates or wages? Minimum wage is $7.25 per hour. A skilled craft worker (pipefitter or electrician for example) might earn four times that amount, around $30 per hour. But an attorney, senior manager, or consultant might charge $500 per hour—nearly 70 times the minimum wage! I recently heard of a life coach who was charging business executives a cool $1,000 per hour and was booked solid!

So, why do some people earn ten bucks an hour while others charge hundreds of dollars per hour? It is simply about the value he or she brings to an organization and skill set that the professional has developed over time. But here is the key for all of us; we can learn at least some of the secrets of $1,000 per hour work and apply them to what we do day in and day out. And, in so doing, change the true value we bring to our teams and the companies in which we work.

About Safety—I'd say the concept of 'ten dollar an hour work versus five hundred dollar an hour work' is very applicable to utility safety professionals and to those who manage safety sensitive work (first line supervisors, managers and directors). In safety, the 'hard' stuff is easy. The hard stuff are things we can feel, touch and put our hands on like the safety rules, procedure book or OSHA regulations. But, the 'soft' stuff is hard. The 'soft' stuff is coaching and feedback, work observations, effective communication. Working on the soft stuff is five hundred dollar an hour work. Safety sensitive environments are rich with these opportunities. Here, five hundred dollar an hour work can drastically change the results.

Why Transition to Five Hundred Dollar Per Hour Work?—Decades ago in Paris, legend holds that a lady was walking through the streets shopping when she noticed a painter who was set up and painting on a street corner. The lady walked over and interrupted the man's work to ask if he would paint her. Without much discussion he stopped his work, pulled out a new canvass and went immediately to work. Within a few minutes he handed her the painting. It was beautiful, exquisite. Pleased with the work, she asked what she needed to pay for the painting. He replied five hundred dollars. Shocked, the lady said, "Well, sir, it only took you a few minutes." To which the painter replied, "No, that has taken an entire lifetime." The painter was Pablo Picasso.

Think In Terms of Blue Chips!—Often, if I am conducting a full day or multi day seminar, I will end the session with the blue chip activity. Two participants will stand with their backs to a table as the rest of the participants gather around to cheer on these two volunteers. When I say the word, they will turn and begin to pick up discs. After about ten seconds I will yell stop. What generally happens is that the participants will turn around and immediate start picking up chips, the ones that are closest to them. What they fail to realize is that there are three different color chips, white, red and blue. When they turn around, the white chips are right in front of them and there are lots of white chips, so they focus their entire time picking out these chips. What they don't realize is that white chips have a value of one, red chips have a value of ten while blue chips have a value of 500! In a recent session, I asked a participant why he didn't pick up blue chips, he said, "I didn't even see blue chips!" He was totally focused just on those chips immediately in front of him.

Each day supervisors and middle managers are pulled in a dozen different directions. An entire day can be largely spent putting out 'the fire of the hour' only to move to the next fire, then the next (picking up the chips that is closest to them). Or, if one is not putting out fires you are likely trying to get caught up on the scores of paper work, phone calls or emails. In both cases, this is the work that is closest to them, white and red chips. All of your work is important, or you wouldn't do it, but some work brings more results than other tasks. The first objective is to understand what is ten dollar an hour and what is five hundred dollar an hour, or a blue chip. Once we have that understanding and can see blue chips in our workday, we simply do more of the five hundred dollar per hour work. Doing this over the next few years will bring you added skills, and value to your organization.

Beyond Coaching Rules—A study entitled "The Peer Principle" by *Bloomberg Businessweek* published in May 2010 stated, "In the area of safety, our study found that 93 percent of employees say they see urgent risks to life and limb, and yet less than one-fourth of those who see concerns speak up about them. Rather, they wait for bosses or others to take action." Researchers compared organizations with average safety performance to organizations with exceptional safety performance. At first, they could not find a difference. But after thousands of interviews with managers and first line supervisors, they found some eye opening information.

In the end, researchers found that accountability was the key element to outstanding safety performance. But, it wasn't supervisors holding workers accountable that propelled organizations to the next level. Instead, it was workers holding each other accountable. "Remarkably, cultures of accountability had little to do with bosses. Rather, it was all about peers." Organizations with cultures of peers coaching peers found remarkable success—and not just in safety. "Those supervisors and managers with the strongest safety records were five times more likely to be ranked in the top 20 percent of their peers in every other area of performance. They were 500 percent more likely to be stars in productivity and efficiency and employee satisfaction and quality, etc."

Today, most organizations have observation programs where supervisors watch employee work and offer coaching. Yet, these are too often limited to easy stuff—the rules. While the safety rules need to be observed and discussed, so do more important elements like peer-to-peer accountability, job planning, job leadership, job communication. These are harder to grade but if you want to move your organization to the next level, and work toward five hundred dollar an hour work, take this step.

Train, do not Educate—To educate simply means that you give a person information. Training means that you give information with the expectation that a new behavior results. Five hundred dollar an hour work means that one sets up training programs (change behavior) instead of education programs.

Years ago, when I was a safety professional supporting nearly 400 linemen, substation technicians and natural gas pipefitters/equipment operators in out-state Missouri, we determined that one of the contributing factors to a number of injuries was poor job planning. In the electric utility industry, OSHA demands a plan before any work begins and mandates a five step process to job planning. In coordination with others, we put together a top-notch interactive program, and then delivered it. About three months later I was back in front of a group I had trained just 90 days earlier but they didn't know the information. They were not using it day in and day out for job planning. I had educated, not trained.

Virginia Tech Professor and noted safety professional Dr. E. Scott Geller says it this way, "Training requires coaching, which means C for Care, O for Observe, A for Analyze, C for Communicate, and H for Help. This requires observation of behavior and appropriate delivery of corrective and supportive feedback." Five hundred dollar an hour work means that processes are set up to ensure a change in behavior after training, or as Dr. Geller says, COACH.

Elevate Organizational Energy—The sad truth is that I have been to some funerals that had more positive energy than safety meetings. Safety meetings are just one example of a task that all too often is simply a 'check the box' activity. Energy is a key to results, positive or negative. Think about your own life. When your energy is low, the quality of your decisions is lower than if you are feeling good and energy is high. When energy is low, you are less tolerant, coach less and offer less feedback compared to when your energy is high. Five hundred dollar an hour work includes evaluating your teams or the organization's energy and finding ways to move that energy to higher levels.

Jim Loehr and Tony Schwartz in their groundbreaking book entitled *The Power of Full Engagement: Managing Energy, Not Time, is the Key to High Performance and Personal Renewal,* said the following. Keep these quotes in mind as you energize your team! "Energy, not time, is the fundamental currency of high performance. Performance, health and happiness are grounded in the skillful management of energy." And, "Leaders are the stewards of organizational energy!"

Each day we have a choice. Are we going to pick up white chips and red chips, or are we going to make sure that we spend more and more time each day with blue chips. If you do, someday your name may be added to the five hundred dollar an hour list. But more importantly, you will be setting safety within your team and organization on a new path for success. The money is nice, but safety is a reward that saves lives!

Sources:
Grenny, Joseph, "The Peer Principle," *Bloomberg Businessweek—The Influential Leader,* May 2010.

Loehr, Jim, *The Power of Full Engagement: Managing Energy, Not Time, is the Key to High Performance and Personal Renewal,* Free Press, 2003.

38 What Day Will You Get Hurt?

It's about an Attitude, Not a Day

Can we statistically determine what day our workers will get hurt? And, can the day of the week that an injury occurs mean that severity will be less, or greater? Over the last few years, there have been some industry experts who have predicted that injury severity increases on the day before a weekend or the day before an extended break. Can a quick Internet search support this theory? Let's take a look. On Saturday, May 8th, a gas explosion in China's Hubei Province mine killed ten and injured six. A weekend. A week later, China experienced another explosion and coal mine disaster, this time it happened late week, on a Thursday. In this case, 21 miners were killed. Do we have a late week trend? On further searching, we find that the BP gulf coast incident happened on a Tuesday. That seems to blow the late week theory. Or, the space shuttle Challenger exploded on a Tuesday as well. This myth might be busted.

In looking at a number of other incidents, injuries, and fatalities, we probably can't determine a statistical probability of an injury happening on a specific day of the week but if we look closer, we can find some themes. And, discovering these themes will allow us to be aware and prepared, ahead of a potential disaster.

End of a Job—The single deadliest event on Mount Everest was on May 10, 1996 when seasoned and experienced guide Scott Fischer and seven other climbers were killed—they were on the decent. At the beginning of a hazardous job, we tend to be on our toes. We plan. We are vigilant. Our awareness is heightened. Yet, once we reach that peak, we tend to think the major hurdles are behind and we can let that guard down a little. *Forbes* associate editor Christopher Helman wrote the following in a recent article about the BP disaster, "We know with some certainty that workers were in the final stages of setting the final sections of pipe (production liner) in the hole and cementing it in place. The plan was to set cement plugs in the well, temporarily abandon it, and move the Deepwater Horizon off to a new drilling site within a couple days." Did the fact that they were 'coming down the mountain contribute to the incident? I'm not sure, but the end of a job or task can mean that we let our guard down allowing injury or disaster to creep in. Take extra precautions; both climbing and descending dangerous tasks.

Change of or Absence in Supervision—"Although it wasn't, May 2, 1972 almost felt like a Friday for the 173 miners reporting to their normal 7 a.m. to 3 p.m. day shift," Matt Forck writes in his new release, *Check Up From the Neck Up: 101 Ways to Get Your Head in the Game of Life*. "The atmosphere at Sunshine Mine in Kellogg, Idaho, probably felt different because the top brass was several counties away attending the annual stockholder's meeting. With the 'big bosses' gone for the day, it seemed that everyone was taking it a step slower." It was early in the shift that the fire alarms rang. A fire in a mine can lead to disaster but this was a silver mine and didn't offer much fuel for a fire. Tom and Ron, partners for the last several years, left their post and headed to the man lift to go topside. On the walk there they joked that this might even mean an early beer at the local pub. Once at the man lift, waiting with dozens of other men, Ron collapsed, overcome by fumes. Tom grabbed him and pulled him back near their work location, to fresh air. Once Ron was feeling better, they again headed to the man lift. What they found there horrified them. All of the other men waiting for the lift, just minutes earlier were joking and laughing, were now dead.

Things change when the boss is out of town. When management shifts, a new boss is hired, one retires or someone is temporarily upgraded to fill a role. Attitudes change when management is off-site at an event, all day meeting or stockholders meeting. These situations can't be avoided but when they occur, be aware of job assignments, crew assignments and production rates. Instruct those leading the work to take extra time planning. It's even a good idea for the management left behind to be active in the field or on the floor, just to make sure work is progressing safely.

By the way, eight days and over 200 hours later, rescue crews reached Ron and Tom.

What Day Will You Get Hurt?

Once safely above ground they learned they were the sole survivors of one of the worse mining incidents in the United States; an incident that took 91 lives.

A Simple and/or Routine Job in Combination with Weekend or Break— It seemed to be an easygoing Thursday morning. It was in a safety committee meeting when my phone rang. I first ignored it, intending to dedicate my energies to the meeting. Yet, the phone rang again, and then again. I stepped out to take the call. It was the regional dispatcher. He told me that we had an electrical contact. He informed me that emergency services, including the life flight helicopter, were on site. I left the meeting and made the 80-minute drive to the location. I found that the crew was on their last day before a three-day break. I also found that the utility crew was working a very simple pole change-out job; one that each of the six men had done, dozens, if not hundreds of times. In the incident, two of the men had been electrocuted; one didn't make it.

While I may not be able to prove it statistically, I believe that there is something to the notion that opportunity for incident severity increases before a holiday or extended break, in this case a three-day break. But, I think that a combination of a break with a simple and routine task is the combination to watch out for. When this combination occurs, take some extra time in the job planning. Make sure the entire crew discusses all hazards and takes the appropriate actions to eliminate each hazard, according to the rules and policies. Finally, stop the work periodically to make sure everyone is still on the same page and rules are being followed.

End of the Day, End of Job Work Pressure—I was on cloud nine! I had just finished my first presentation at a national safety conference. Since I wasn't flying out until the next morning, I walked the Baltimore Harbor waterfront, located a terrific seafood restaurant and was seated at a window, so I could watch the boats bouncing in the harbor. I was somewhere in the middle of my salad when the cell phone rang. It was a good friend, a safety professional with a utility, and he needed someone to talk to. He told me about a utility incident that happened just hours earlier. A service worker, at the end of his shift, was asked to install some labeling in a piece of energized electrical equipment. Although the job was very simple, the service worker apparently hurried to complete it. In the process, he contacted energized high voltage equipment. He was in a burn unit, clinging to life.

We feel pressure to hurry, to get it done. And this pressure is never greater than at the end of a shift. When we are racing to complete a job near shifts' end, take a few seconds to stop and perform a safety stop. A safety stop is when the entire crew stops, reviews the work and the safety work rules associated with the task, and then continues. This 90-second safety stop can literally save lives at the end of the day.

Remember, maybe the most famous end of shift work pressure incident was the space shuttle Challenger disaster. As you remember, the Challenger splintered into millions of pieces when it blew up 73 seconds after lift off. To meet a pressure packed deadline, the decision was made to launch after some engineers questioned how an O-ring seal in its right solid rocket booster would respond in the cold weather. If there wasn't pressure of the 'deadline' would that decision have been different?

I'm not sure we can statistically prove that injuries or incidents will happen on specific days or at specific times, yet there are some warning signs to look out for. I think that there is a tendency to let one's guard down after the 'heavy lifting' on a job is finished and we are coming down the mountain. I think that when supervision shifts, we have simple tasks before a long break or we are hurrying to finish a job before the shift ends, all present certain dynamics that can lead to an incident. As safety leaders we need to be ready and aware that certain conditions make it easier for an injury or incident to occur. Where are those conditions in your work environment? And, what are we prepared to do to prevent 'bad' stuff from happening?

39 How Can Safety Leaders Sleep at Night?

They Know These Seven Keys to Worker Engagement!

Gallop defines engaged employees as "having 100 percent psychological commitment to their job." Given that definition, one of my favorite one-liners about engagement reads, "He is about as engaged as a kamikaze pilot on his 18th mission!"

The fact is that our team's engagement will determine how quickly we achieve results (meet our deadline, hit that goal) and the quality of those results (completing the job safely). Results are directly proportional to the level of engagement.

Given how important engagement is to results, knowing these seven key points allows utility leaders to sleep at night—and get results.

1. Engagement Is a Big Deal—When it comes to unengaged workers, the data is sometimes overwhelming. One report estimates that disengaged workers cost the U.S. economy $350 billion a year in lost productivity. Worker engagement can't be easily gained or even 'bought' through improved benefits. New research by Harter and Sangeeta Agrawal, a Gallup research manager, shows that employee engagement has more of an affect on employee wellbeing than vacation or flextime policies do. Simply said, engaged workers are happier.

But, just as unengaged workers hurt the bottom line, engaged workers make a dramatic improvement to that bottom line. A recent study in the health care industry found, "Those who are engaged are safer, more productive, more likely to stay with the organization, and more likely to provide outstanding patient care. For example, employee engagement and safety climate were inversely correlated."

2. Define and Policy Up—"People play harder when they know the score," anonymous. When it comes to engagement, knowing the score is having a firm definition of exactly what engagement means.

To find the score, Jennifer Robinson in her insightful article; "For Employee Wellbeing, Engagement Trumps Time Off" defined the three types of employees. There are employees who are not engaged, "employees are essentially 'checked out.' They're sleepwalking through their workday, putting in time—but not energy or passion—into their work." She defines actively disengaged employees as "employees that aren't just unhappy at work; they're busy acting out their unhappiness. Everyday these workers undermine what their engaged coworkers accomplish. Finally, she details engaged employee, "who work with passion and feel a profound connection to their company. They drive innovation and move the organization forward."

There are complex ways to measure engagement, such as annual employee surveys or consultant studies. But, these definitions are easy, and can allow any supervisor or manager to take the score any time he/she needs to.

Next, I'd bet that your organization has step-by-step procedures on all important safety sensitive tasks. I call these the hard skills. The reason they are 'hard skill's is that OSHA, your industry, engineering practices or other disciplines have set out hard facts and rules or step by step guides for these tasks. Yet, these hard skills are only half of the equation for safety results. There are a number of people skills, also called 'soft skills,' such as engagement, leadership, coaching and feedback, etc. Each of these soft skills also needs a step-by-step 'how to' outline, measurement tool and accountability element to ensure their success. Would you or one of your employees enter a confined space without a written procedure? Why is engagement any different?

3. The Goal Is Peer-to-Peer Feedback—A study entitled "The Peer Principle" by *Bloomberg Businessweek* published in May 2010 stated, "In the area of safety, our study found that 93 percent of employees say they see urgent risks to life and limb, and yet less than one-fourth of those who see concerns speak up about them. Rather, they wait for bosses or others to take action." The 'bosses can't be everywhere, nor do we want him to be. Instead, one of the goals of engagement is to foster peer-to-peer feedback. Successful companies find that workers talk to workers, they don't wait on the boss.

Remarkably, cultures where peer-to-peer feedback, also called accountability, was the standard practice found this, "Organizations with cultures of peers coaching peers found remarkable success—and not just in safety. Those supervisors and managers with the strongest safety records were five times more likely to be ranked in the top 20 percent of their peers in every other area of performance. They were 500 percent more likely to be stars in productivity and efficiency and employee satisfaction and quality, etc." Can you say, engaged workers!

4. Give Your Workers the 'Why'—Jason Milton in his article "Great Leaders Inspire Excellence Through Mission" writes, "For years, a quiet orderly mopped floors, scrubbed patient rooms, and cleaned up bodily fluids. Never complaining, always smiling. At the end of one of his graveyard shifts, a cynical colleague asked him how he could be so happy about mopping the same floors every day. Puzzled, the orderly replied, 'I've never mopped a floor a day in my life. I work here so I can stop deadly diseases from infecting others.'" The ever popular Mission and Vision statements have often failed to engage our workers in the greater purpose, the 'why.' For this orderly, his why was clear, he was engaged because he saved lives by stopping deadly disease. Your company might make a widget, but chances are that widget is part of a larger whole that greatly contributes to humanity. Find your why and engage your people with it. Even the most unengaged worker will raise an eyebrow about saving a life. Milton writes, "The best leaders make their company's overall mission clear and help employees understand how their work contributes to a greater purpose."

5. Make Caring Personal—Tony LaRusa, the third most winning major league baseball manager says that leadership must be personalized. Making leadership personal to each player leads to engagement, and results.

When I was a safety supervisor for a major mid-west utility, I served over 400 line workers, substation technicians and heavy equipment operators. With so many people in over twenty locations, there was no way I could see each worker each week, or even each month. So I set up a number of ways to communicate, one way was the personal birthday note. For two straight years, I sent each worker a personal handwritten note on his or her birthday. Some years later, I was in a workgroup and saw one of my notes on the outside of a locker. I found the worker, he was a thirty year veteran who some would have labeled as 'trouble' and asked him why he taped his birthday note to the outside of his locker. He said, "In thirty years of working here, no one has ever taken the time to write me a note, and say something so kind. I taped it outside my locker so I can see it every day."

Eric Jackson writes, "Sure, people come to work for a paycheck, but that's not the only reason. In fact, many studies show it's not even the most important reason. If you fail to care about people at a human level, at an emotional level, they'll eventually leave you regardless of how much you pay them. Many people assume that it's possible for a person to be an effective leader without being likable. That is technically true, but you may not like the odds. In a study of 51,836 leaders, we found just 27 who were rated at the bottom quartile in terms of likability but in the top quartile in terms of overall leadership effectiveness—that's approximately one out of 2,000."

When the veteran worker talked, I heard, no one had ever tried to engage me on a personal level in three decades. And given how he responded to just one simple note, I believe that consistent engagement would have easily moved him from 'trouble' to contributor. Make engagement personal, and your personal challenge.

Leaders need to know their stuff, and for today's leaders, their people need to know they personally care.

6. Engagement Is Money!—Captain Mike Abrashoff took command of USS Benfold, of one of the worst performing ships in the Pacific fleet, and led it to the top of all measurable categories in less than two years. He retells a number of powerful stories in his best-selling book entailed *Its Your Ship, Management Techniques from the Best Damn Ship in the Navy*. One of my favorite ones, which really showed the power of engagement, is when his ship needed to be recertified for battle readiness. At that time, certification was based on passing a week-long series of written and skills based exam. To prepare, the Navy ordered all ships to six months of battle ready training at sea. Due to his crew's engagement, they had revamped the training practices, found ways to foster lower preforming sailors, challenged high performers and spent time on realistic battle scenarios versus mindless training. They tested before launching for the six month sea-based training and passed the test—okay, they aced the test. They achieved the highest test score ever recorded, including ships that had tested for six months at sea!

Abrashoff called his commanding officer, delivered the news and explained how they would not have to go to sea for six months. There was a long pause on the line, and his Commanding officer replied, "We're not set up to do that, you have to go to sea for six months." Abrashoff then explained how much money could be saved by not going to sea and it was agreed, the crew of Benfold could make other training and educational arrangements. When I read, 'we're not set up to do that,' . . . I read, 'we are not set up for success.' Engagement can lead to success, and also can save you a lot of money!

7. The Buck Stops at the First Line's Desk—The complaint from most first line supervisors is that everything, and I mean everything, gets dumped on them. Having sat in that chair for two years, supervising electrical linemen, early in my career, I agree that every good and not so good initiative lands on the first line's desk. But, when it comes to engagement the data seems clear. Gallup, from decades of research, reports that employee engagement, which is highly correlated with productivity and the company's market value, will soar or plummet depending on the employee's relationship with their manager.

Eric Jackson in an article entitled "Top Ten Reasons Why Large Companies Fail To Keep Their Best Talent" writes, "So, for all those employers who *have everything under control*, you better start re-evaluating. There is an old saying that goes; "Employees don't quit working for companies, they quit working for their bosses." Regardless of tenure, position, title, etc., employees who voluntarily leave, generally do so out of some type of perceived disconnect with leadership.

Here's the thing, employees who are challenged, engaged, valued, and rewarded (emotionally, intellectually & financially) rarely leave, and more importantly, they perform at very high levels."

Sure, first lines have a lot to do, but engagement and the tactics and leadership needed to build upon it is not just something else for a first line to do, it is what a first line needs to do before all else. The buck stops with engagement, and with the first lines to engage.

Sources:

Roberts, D. Steve, and E. Scott Geller, *An Actively Caring Model for Occupational Safety: A Field Test,* 1995.

Thorp, Jonathon, MD, MBA, Waheed Baqai, MPH, Dan Witters, MS, Jim Harter, PhD, *Workplace Engagement and Workers' Compensation Claims as Predictors for Patient Safety Culture,* December 2012.

Nink, Marco, *German Workers Equally Satisfied With Male and Female Managers,* The Gallup Blog, September 2012.

Milton, Jason, *Great Leaders Inspire Excellence Through Mission,* Gallup, July 2012

Robison, Jennifer, "For Employee Wellbeing, Engagement Trumps Time Off," *Gallup Business Journal,* December 2012.

Dishman, Lydia, "Secrets Of America's Happiest Companies," *Fast Company Magazine,* January 2013.

Jackson, Eric, "Top Ten Reasons Why Large Companies Fail To Keep Their Best Talent, " *Forbes Magazine,* December 2011.

Grenny, Joseph, "The Peer Principle," *Bloomberg Businessweek—The Influential Leader,* May 2010.

40 Yield, For Safety's Sake

There is a mind-set in safety sensitive tasks, nothing bad ever happens if one just stops and reassesses. Or, to reword—yield, for safety's sake.

The day was really like any other; it started better than most. It was Thursday (I like Thursdays). I had breakfast with the family. I kissed them off to school (my kids are in 6th and 7th grade and my wife is a 5th grade teacher at the same school), and then I started the morning drive.

I rolled into McDonalds for a one-buck tea to find that there was absolutely no one in the drive thru line—my good day was looking even better. Back on the highway, I was on cruse control thinking of the day ahead. Out of the corner of my eye I saw a white car entering the highway at a fast pace. In my quick assessment the driver of this car seemed intent to go from the east side of a four lane highway to the west outer road. On this particular stretch of highway there wasn't an on or off ramp; one crosses over the north bound lanes, yields in the middle, then proceeds across the south bound traffic.

The fact the car was moving fast caught my eye. She raced across the north bound. She yielded, barely, and raced into the south bound lanes. I believe she eyed the two cars in the south bound passing lanes and was intent on beating them. She did. But, in focusing on getting ahead of them, she didn't see a black Ford Expedition in the south bound driving lane. The impact still burns in my mind.

I was one of the first cars behind the crash. I pulled over. I called emergency services. I got out of my car to see how bad it was. Steam and smoke misted from what was left of the engine compartment. I was joined near the white car by four to six others a few wearing scrubs—that was a clear sign they knew more about medical aid than me. The women in the white car was laying lifeless across the front seats. One person checked a pulse and heart rate. She had both. I yielded . . . and prayed.

About a decade ago I remember teaching a defensive driving refresher course to utility workers. In that training, there was a concept called a safety stop. When a utility worker approached an intersection and decided to execute a right turn on red he would first make a full stop. Once stopped he would check the entire intersection, left, straight and right. If it was clear he could proceed he would slowly begin to execute the right turn. But, before entering the intersection, stop one more time—a safety stop. Here, the driver checks one more time to make sure the intersection is clear, and if it is he may proceed. This stop is synonymous with a yield.

In safety we need to yield.

Yield to hazards, making sure we have mitigated all risks before moving ahead.

Yield to safety rules, ensuring compliance with all rules and policies.

Yield to your families and friends, as they are depending on you to make safe choices and return home whole.

Yield to your hopes, dreams goals and ambitions; for those can only come to fruition if you are healthy and intact.

Later that same day a friend who I had told about this horrific crash sent me the following news article. It read, *"One woman was killed and another seriously injured in a Thursday morning crash on Highway 63 near Ashland. A 19-year-old driver was crossing Highway 63 from Forsee Road in a Dodge Avenger and failed to yield to 34-year-old Trudy who was driving a Ford Expedition who was driving south on Highway 63. The SUV hit the car on the passenger side. The 19-year-old driver was pronounced dead at University Hospital."*

Yield for your safety today. The 19-year-old's friends and family wish that she had. And please, keep them in your prayers.

41 Seven Strategies for Improving the Safety Record of Your Crew

Greg, a lineman from Chicago, wanted something different for his family, so he moved from the Windy City to a small rural town in northwest Missouri. Instead of a show up location with more than 200 linemen, his new show up had 15 linemen, a storekeeper and a supervisor.

One of the first things he did was bring in his son, Ben, to work. Ben was able to see his father's tools, meet his dad's coworkers and boss, and even go for a ride in the bucket truck. This fun afternoon made Greg certain that he did the right thing by moving his family to small town America.

Driving home that afternoon, Ben looked at his dad and told him, "Dad, I don't like you being a lineman."

"Why?" Greg asked.

"Well, Dad, it looks dangerous," he replied.

"Well," Greg responded, "it is, but as long as I don't take any shortcuts, nothing bad can hurt me."

Greg told me that this must have been enough for Ben, because he didn't say another word. Months passed. Then one day, the utility notified Greg that he and his crew would respond to Hurricane Dennis. After leaving, he found a surprise note in his lunch box, and when he returned, he was kind enough to share it with me. It read, "To Daddy, I love you. I will really, really miss you a lot. Don't take shortcuts. Love, Ben."

This story reminds us of the importance of bringing home lives into our work lives. We can increase our energy, focus and safety results by trying the following seven strategies.

1. Personalize the Work space—Position a bulletin board near the locker room or in an area close to where your linemen begin their day. Invite all workers to bring in pictures from home. Remember to update the board frequently.

2. Create a Strong Bond Between Linemen's Families—I love the old saying, "The family that eats together stays together." I strongly believe the work group that eats together is safer together. Set up two to four family events for your work group. Think about some basics like a summer family picnic, a football-themed chili cook-off, a holiday party and a Memorial Day barbecue. Whatever the schedule, the important thing is that families get together and become closer, and ultimately, it will bleed over into work life and reduce shortcuts.

3. Make a Token Reminder—I have worked with some utilities that offer a special coin for safety. By giving this tangible item to their linemen, they remind their work crews to focus on going that extra step to customize each coin with the name of each family member.

4. Send Letters—The letter that Ben sent his father had a great impact on him. What if your employees received a letter from a loved one urging him or her to work safely? I am sure it would have a similar impact. To that end, organize a letter-writing campaign for your employees' children and loved ones. Use the emergency contact numbers as a starting point. I have helped a few work groups facilitate this activity, and it has been very successful.

5. Sponsor a Safety Fair—In our line of work, it's really hard to bring our children to work; there are simply too many hazards to do so. But, we can pick one Saturday each year to hold a Safety Day. We can set up poles for climbing, meters for installing and bucket trucks for riding. This is a great opportunity to let our families know

Seven Strategies for Improving the Safety Record of Your Crew

what we do at work each day, and for them to give us feedback about working safely.

6. Organize a Safety Contest for the Kids—Many organizations are now offering an annual safety poster contest for the children and grandchildren of employees. This gives your children and grandchildren an opportunity to tell you why you need to work safe.

7. Move Safety Home—As we know, if one of our linemen gets hurt at home, it still affects us at work. At-home injuries may not be OSHA recordable, but they still have an impact on the work environment. If one of our coworkers gets hurt at home, he or she may not be able to make it into work. To prevent this from happening, utilities can offer a safety recognition program. Linemen can earn points or credits for certain at-home activities such as changing their smoke detector batteries. After they earn a predetermined amount of points, then they are eligible for a prize or a family centered gift card.

One of my favorite sayings reads, "No one told us we had to take the fun out of work, we did that all on our own." In the same vein, we don't need to take family out of work. If we can find a way to keep them with us at work, we will see it in our safety results.

42 Constant Safety Awareness

Five Keys to Managing Your Space Between

Anyone who has ever climbed a pole or opened a switch knows about safety. Linemen wear rubber gloves, hard hats and steel-toed boots. Apprentices learn the correct way to do the work. These safety values and skills are slowly absorbed as hundreds of hours are spent working alongside journeymen linemen in storms, on night calls, and in routine maintenance and construction assignments. When linemen top out as journeymen, they are trained, skilled and knowledgeable to do their work safely. They know the right way to perform each and every job.

Linemen are physically fit and mentally tough, too. They spend their days in all types of conditions, from extreme heat to sleet and snow, and from hot dry pavement to wet and slippery mud. If push comes to shove, they can hoist a 50-kVA tub with a couple of men and a set of blocks. The question remains: Why do men and women in the line trade still get hurts and unfortunately even killed?

Reflecting on my own career and incidents that I had witnessed as a journeymen lineman, a supervisor and safety professional for a Midwestern utility, I found a pattern. I discovered a space between what we know is right and the choice we sometimes make. Unfortunately, these choices sometimes lead to mistakes, incidents and injuries. The following are five keys to managing your space between to ensure you and your coworkers go home safe each day.

1. Use Tools in the Proper Manner—A couple of years ago, a line worker was in the air drilling holes to frame a pole. A strong wind blew wood chips right into his face. Although he called for additional eye protection, which was in the truck bin, he continued drilling, using his hand as a shield instead. The crew spent the rest of the afternoon in the emergency room flushing his eyes.

If linemen know anything, it's how to improvise, which can be both a strength and a weakness. Improvising with tools, or stopping short of using the right tool for the job, leads to incidents and injury. It's just a matter of time.

2. The Little Things Matter—Who really needs a wheel chock? One day a crew was unloading a backyard machine from a trailer hooked to a one-ton pickup. Next to the trailer was a piece of wood, which someone kicked under the trailer as a wheel chock. As the linemen walked the yard machine off the trailer, the weight shifted and the whole rig slid down the hill. Everyone scattered like a flock of birds as the unit sailed through a yard and rested against a tree. Fortunately, no one was hurt by the runaway equipment, but it could have led to a serious injury for a crew member or the public. By the way, the wheel chock was on the trailer, but the crew opted not to use it. Little things, as you know, can make a big difference.

3. Take the Time It Takes—Since some linemen are in the air all day every day, being 30 or 40 feet in the air may seem like standing on the ground. I remember one time a lineman was working in a substation and needed to remove a set of grounds. He couldn't break the ground loose with the shotgun stick, so he climbed off of the ladder onto the top of a transformer.

He was about 12 feet in the air and didn't take the 10 seconds to go back down and get his harness. He thought he didn't need fall protection, because he would only be at the high elevation for a short time. With leather gloves, he used a screwdriver to break the ground lead free. When he removed the ground, he was immediately shocked by induction voltage. It wasn't enough to kill him, but it knocked him off the transformer. He fell head first, his body hit the ground and everything went black. When he finally woke up, he spent several months in the hospital and endured several surgeries. He was fortunate enough to make a recovery, but others aren't quite so lucky. Safety only takes seconds, but injuries last forever.

4. Wear Personal Protective Equipment—Along with fall protection, linemen may believe personal protective equipment (PPE) is optional, but it's not. For example, when hooking up a service, rules require that a lineman wear a hard hat, safety glasses, a harness, and low-voltage rubber gloves. One day, I stopped one of my troublemen, who wasn't wearing any of his PPE, except eye protection. I called him down and asked why.

"Son," he told me, "I've been doing this since before you were born. I know what to do and when to do it."

Many linemen have developed a bulletproof mentality that an accident can't happen to them and PPE is not needed. Since we cannot predict when and where an incident can happen, PPE should always be mandatory.

5. Tailgate—Every day, call the crew together, identify what hazards are out there, what can hurt you and what safety rules apply to the work at hand. In addition, stop throughout the day to make sure everyone is still on the same page. I remember when I was an apprentice, an old salty foreman would call everyone together every day to discuss job hazards. We would review the job, identify hazards and discuss the right way to get the work done. At the end of the tailgate, he would add, "And one last thing, no one gets hurt today." That still echoes in my head: "No one gets hurt today."

Linemen work around high-voltage lines, dangerous heights and extreme hazards, yet sometimes fall into the trap of thinking they're immune to injury. By taking shortcuts, however, linemen can put their lives on the line. The basic safety rules will help linemen protect themselves and their coworkers in the field and make the line trade a safer occupation for everyone.

43 Five Hidden Safety Secrets of Line Work

In the transmission and distribution world, we don't often get back our lost opportunities. Our work is unforgiving.

One safety mistake is life changing, not only for the one making the mistake, but also for his or her family and coworkers. Yet, within our experiences and those of our coworkers, there are hidden secrets to safe work. The key is to find them and use them. Below are five hidden secrets that I have discovered.

Safety Secret 1: Always Ground Equipment—Unfortunately, in the utility business, workers are injured and killed when equipment contacts distribution lines and linemen on the ground are in contact with that equipment. Years ago, I was the lead on such an incident, analyzing a fatality that resulted when a worker received a fatal electrical contact from a piece of equipment.

The equipment had contacted a distribution line while a worker was in contact with the equipment.

While reviewing the incident, we studied equipment grounding practices and read white papers written by the leading experts. One of the engineers wrote, that in his experience, he has never known of a fatality resulting when a grounded piece of equipment contacted a line. The point is that linemen shouldn't debate whether or not to ground the equipment. Instead, they should know their rules on proper grounding and just do it.

Safety Secret 2: You Pay When You Get Hurt—A couple of years ago, the Centers for Disease Control (CDC) released one of the more comprehensive studies on the dollars associated with one year of personal injuries. The report revealed workplace injuries occurring in 2000 cost a combined $505 billion.

Yet, the true take-away from that report was the personal financial impact of a workplace injury on an employee. The CDC looked at the traditional costs paid by the employer such as medical costs, physical therapy and workers compensation. In addition, for the first time ever, they were able to put a dollar amount to the loss of wages, a reduction in fringe benefits and standard of living due to an injury. These are costs that each individual person and his or her family will shoulder. In the end, the study showed that for every dollar an employer pays for an injury, the worker that was hurt will pay three dollars, and that's a secret worth remembering.

Safety Secret 3: Be an Educated Lineman—Several years ago, when I was a very young journey line worker, I was working with a seasoned crew leader and an apprentice. We were terminating URD cable in a new subdivision and took a break. While on break, the old foreman took the cable cleaner we had been using as part of cable terminations, and began to us it to clean a hot stick.

I had just finished a week-long school on undergrounding where I learned about a number of things including the fact that approved cleaners needed to be used only for approved applications. I also knew that we had approved hot stick cleaners. Add to that the simple fact that we put our lives in the integrity of those sticks since we used them to move and operate wires and hardware energized up 100,000 volts.

So I voiced my opinion. The simple fact that the cable cleaner should not be used for hot sticks.

The crew leader looked at me, threw the can in the ditch and said, "That's all we need now, educated lineman!" The truth is that our work is changing every day, and the best thing we can do for ourselves and for our safety is to be educated linemen.

Safety Secret 4: Know the Rules—Shortly after I was promoted into a safety role for a utility, I was performing some job inspections, and I pulled up on a crew working on some 34.5 kV. The six linemen on the job had more than 150 years of combined experience. I noticed, however, that something was wrong with how the line was grounded. We stopped work and huddled up.

I pointed out the fact that the line wasn't grounded according to the rules. They looked at me like I was nuts, agreeing simply, "This is the way we have always done it."

Five Hidden Safety Secrets of Line Work

Over the tailgate of a truck, we opened the rule book and learned the proper way to ground the line. When the crew was returning to work, one of the guys asked me if this was a new rule. We looked back at the rule book and found that the section had not been revised for two decades.

Safety Secret 5: Fight for Your Own Safety—This scenario, or something like it is repeated across the utility industry hundreds of times each day. A safety supervisor will show up on a job and discover a few issues that are not right. It could be something thought to be minor, such as a lineman not wearing safety glasses, or it could be a major violation, such as inappropriate cover up. In either case, the safety supervisor will stop the job, call the crew together and explain the rule violation and why it's important. The crew members will then tell the safety supervisor why they cannot comply. Rather than arguing against your own safety, be responsible for it. If someone is trying to give you feedback, don't fight it because the person is trying to help you.

In the end, grounding, the true cost of injuries, education, knowing the rules and fighting for your safety are all key elements to safety success. They are good reminders, hidden secrets if you will, to safe and productive work.

44 The S.T.O.R.M. Model for Near Miss Reporting

Near miss events cost our industry twice as much per year than fatalities. And, these are time bombs waiting to hurt someone if corrective action is not taken.

The non-reporting of near miss events happens for a variety of reasons, ranging from lack of motivation to internal bureaucracy. That being said, from time to time, there is a model in the utility industry that seems to work well.

For example, line crews seem to do a good job of communicating during a storm-restoration effort. After a storm, weather conditions are generally poor, electrical hazards are extreme and work shifts are insanely long. To help mitigate these hazards and maintain a high level of safety awareness, crews will gather every morning to review hazards and then meet again in the evening to review the day.

During these evening sessions, line workers will openly share near miss events as a way to keep their coworkers out of danger. Yet, after a storm, most of this free sharing is lost as crews revert back to their normal routine.

The following S.T.O.R.M. model can help line crews to identify near misses, and eliminate the hazards and unsafe conditions that lead to them.

1. See the Near Miss—One of the most effective football coaches of all time, Vince Lombardi, used to say, "It's hard to be effective when you are confused!" That same spirit translates into near miss reporting. The fact is that linemen have a hard time defining a near miss and have an even harder time translating a field event into a near miss report. And, when there is uncertainty, people will withdraw into silence and status quo.

2. Tell the Group at the Next Daily Briefing—Make it clear that any incident, strange happening or vital piece of information can safety be shared at the daily briefing. These meetings should be short and held in a different location than your typical safety meetings.

It's a good idea to begin each meeting with a meeting starter, such as a safety-related news story, a quote or some other short safety-related piece. After the introduction, invite the group to share any near miss, incident or safety insight from the previous day or shift. Then point out any specific hazards for the group such as weather, change in field conditions, or alterations to a specific project. Then, end the meeting and begin safe work.

3. Own the Hazard in Order to Eliminate It—In this step, analyze each near miss that was reported and determine any and all follow up needed to eliminate the hazard. It's often a good idea to go back to the person who made the report, and then work together to make an action plan for future mitigation.

4. Review and Remind—During the daily job briefings, it is important to review follow-up action steps taken as a result of incidents shared during the daily briefings. It is also critical to remind the group of hazards reported and the safety rules and procedures needed to safely deal with such hazards.

5. Move On—The S.T.O.R.M. model is designed so linemen don't have to deal with a lot of red tape or paperwork. Instead, linemen can share a free exchange of information, and then management, safety leaders and workers can discuss the issue in follow-up meetings. After a problem has quickly been resolved, however, linemen must move forward. That way, they can focus on immediate hazards and near miss events.

Storms will come, and storms will go. Near miss events left unattended, however, will continue until someone is finally injured and the situation is addressed. Providing your group with the leadership that allows for swift and effective sharing of near miss events and mitigation of the hazards is key not only to credibility and trust, but also to long-term safety success.

45 Sounds of Silence
What's Wrong with Near Miss Reporting

There is no doubt that near miss reporting is important—very important. As safety professionals and leaders, we know that a free lesson learned today is an injury avoided tomorrow. And there are dollars associated with near misses, serious dollars.

If we ignore the tremendous "human" and emotional impact of serious injuries and fatalities and only compare their financial costs with the costs of near miss injuries, it might be an eye-opening exercise.

According to some estimates, near miss events may cost twice as much as serious incidents or fatalities. According to a Houston Business Bureau, CII and Exxon Chemical report, a near miss event is estimated to cost about $1,300 and they estimate about 1,000 near miss events for every fatality. Using 2004 Bureau of Labor Statistics data, 5,703 workplace fatalities were reported across the United States. At an estimated million dollars per fatality, near misses cost the private sector more than a trillion dollars more, on the monetary side of the equation, than fatality. And, if estimated costs are shifted, near misses can move this ledger amount to twice that of fatalities.

As safety professionals, we already agree about the importance of near miss reporting. Add to this foundational understanding a cost basis rational, then a question flows naturally, why don't we encourage reporting and thoroughly analyze near miss events? After all, how many near misses have you worked on this week, or this month? Chances are, you can count them on one hand—one finger probably. In truth, there are five key reasons why these events are not reported and analyzed. Five truths that no one wants to talk about, until now.

Unspoken Truth I—Near Miss Reports Come with a Cost

Any article or training class dealing with near miss reporting will frame a near miss event as a "free lesson" or a "golden nugget" that must be analyzed for learning. The truth, however, is that these events are not free.

If you are still skeptical, then tell me how many near miss reports have come across your desk in the last week . . . I rest my case. In most organizations there is a cost to near miss reporting. That cost can come in many forms such as loss of credibility by the worker reporting the near miss. The price might be in an intimidating reporting system or unspoken signals from line-management. The cost might be in the perceived time and hassle to make the report. Our critical role is to identify these costs and work to reduce them.

Questioning Attitude—What are the hidden costs to near miss reporting within your organization? How can we put ourselves in the shoes of our workers to truly understand these costs?

Unspoken Truth II—Reporting Requires Motivation

The next truth that no one talks about is this: free isn't enough. There must be an incentive or motivator in order for our people to engage in a near miss reporting process and work an incident through the process.

Put yourself in the place of your workers. It's a typical afternoon, and things seem to be going fine. All of the sudden, bam. The worker is almost injured due to a part defect. She quickly replaces the defective part and begins the task. As she does this she realizes that she was lucky. Had she not noticed the defect there is a great chance she would have been injured. At that moment, the worker thinks "I wonder if I should report it?" The thought that comes next is key—will she be motivated to do so or not? In most cases, "free" isn't enough, there must be a positive motivator established so our workers will answer the "should I report it" question with a strong positive.

Questioning Attitude—Is there motivation to report near miss events? At the moment a worker realizes a near miss occurred, what can be the motivating factor to allow him/her to report the event? What is the "right" type of motivator for your organization? Should I overcompensate with a motivator in the short term to establish a long-term habit?

Unspoken Truth III—Reporting Requires Too Much Work

If we refer back to our college days and reference the great H. W. Heinrich's accident triangle from his book *Industrial Accident Prevention: A Scientific Approach*, we recall that for every major injury there are 29 minor injuries and 300 near miss events. In looking at these ratios in your organization, it is probably safe to say that if reported, a line-manager could have at least one, if not more, near miss events a week.

That being said, how busy are your line-mangers right now? Chances are they are swamped. What are they going to do with another one or two near miss report events per week? Chances are they won't do anything with them, and that's the fear.

In all honesty, what would your organization think of a supervisor who managed over 100 near miss events in a year? While safety professionals would think its great, do you think managers would perceive an issue with the supervisor's performance?

Questioning Attitude—What kind of near miss reporting system can speed up the process, be effective in eliminating future exposures and assist line-managers at the same time? How can line-managers be rewarded and motivated to follow near miss events through the process? How can upper management recognize a high number of near miss reports as "effective supervision" instead of "poor performance" by a line manager? Does an organization need a separate position just to manage near miss events?

Unspoken Truth IV—A Near Miss Is Hard to Define

One of the most effective football coaches of all time, Vince Lombardi, used to say, "It's hard to be effective when you are confused!" The fact is that our people have a hard time defining a near miss and have an even harder time translating a field event to a near miss report. And when there is uncertainty, people will withdraw into silence and status quo.

Questioning Attitude—How can we design a near miss definition that is both effective and easily understood? How can we evaluate our employee's current understanding of near misses?

Unspoken Truth V—Hours of Pointless Questions

Think about it: something happens and the result is that no one gets hurt, yet the employee or crew is dragged into a room with management for an "investigation." After a couple hours of seemingly pointless questions that meeting ends, few, if any, changes are made—what doesn't stink about that?

Questioning Attitude—How can technology be used to help the process, how about a near miss blog? How can the process be turned from "stink" to roses? What if employees ran the process with a fostering eye and consistent support from management?

A Sharing Process that Works

Having worked more than two decades in the utility industry, it is safe to say that the most dangerous work for utilities workers is during storm restoration. After a storm, weather conditions are generally poor and electrical hazards extreme. To help mitigate these hazards and maintain a high level of safety awareness, crews will gather every morning to review hazards and then meet again in the evening to review the day.

It's in these evening sessions that line workers will openly share near miss events. It's a free exchange motivated by a general intent to keep a "brother or sister" from being injured by the same or similar exposure.

The process is supported by line-management and safety staff alike, it is free, workers are motivated by a sense of genuine caring and a desire to help, it's not full of process or forms and it is done in a timely manner. Yet, after a storm, most of this free sharing is lost as crews revert back to "normal" hazards and office politics.

Near miss reporting is one key to safety success, yet it is lost in most organizations. In the end, it's up to us as safety leaders and professionals to speak the unspoken truths and make a process where experiences are shared and supported for one common goal: to eliminate that human side of loss-before it happens.

46 Will You Be Ready When 'IT' Happens?

"He's drowning," students yelled. "Help, someone help him!" shouted teenagers as their fellow classmate KeAir Swift, a 14-year-old freshman, rested lifelessly on the bottom of the pool.

Jonathan Sails, a teacher at East Detroit High School, was not close to the pool, instead he was on the bleachers with other students while class participants in the remedial swim class were in the pool.

Sails heard the shouts and headed over to look in the pool. But Sails didn't jump in to assist. Instead Sails went to the locker room. Sails, the swim teacher, didn't have on swim attire. His trip to the locker room was to suit up, so he could jump in the water. But, Sails didn't have to get wet, the vice principal happened upon the scene and in full street clothes jumped in the pool to get Swift to safety. It was too late. He had been under the water too long. He didn't make it.

Sails was not a full-time teacher, instead a substitute teacher filling in because the 'regular' swim teacher's swim instructor credentials had expired. The East Detroit Public Schools used a third-party contractor, Professional Educational Services Group, one of the largest staffing companies of its kind in the country, to fill substitute positions. So, technically Sails was their employee, not one of the public schools. And, Sails lied about his certifications.

"The defendant misrepresented that he was certified through the Red Cross as a lifeguard, which is what you need pursuant to state regulations to teach a swimming class," says Macomb County Prosecutor Eric Smith.

Unfortunately, this story is true. For the family of KeAir Swift, they wish it was all 'made up' and their son would just return home. "It's terrible and they really need to pay for it," said LaKisha Swift, KeAir's grieving mom. "They really need to pay for it. They need to suffer like I'm suffering. And my son's lost his life. He had a great life. He was a happy child."

Jonathan Sails now faces charges of involuntary manslaughter, which carries the potential of a 15-year sentence and/or $7,500 fine.

In utility safety there is a thought that says, "Where will you be when 'it' happens." The meaning behind that saying is that all things mechanical can, and will, fail. And, all people make mistakes. So, 'it' will happen, just like it did in this remedial swim class, for KeAir, for East Detroit Public Schools, and for Mr. Sails. In utility safety, if we are following the rules, executing our work plan, 'it' will simply be a near miss, and something we talk about and learn from for years to come. But, if we are not following the rules—bad things will happen when 'it' happens. Just ask LaKisha Swift.

After the incident, East Detroit Public Schools released this statement, "The teacher being charged is an employee of an outside contractor that has provided substitute teachers for the district for several years." It did not name the company. Superintendent Joanne Lelekatch followed up the statement with this comment, "When tragedies such as this happen, our focus remains on the student, his peers, and the family."

Be in compliance when 'it' happens, and keep the Swift family in your thoughts.

47 Random Acts of Safety Kindness

In January 1945 the Red Army was closing in on German held territories Within those German occupied areas were a number of concentration camps, including Auschwitz. The German's, wanting to keep these prisoners for labor, decided to move them; they were told to march.

These prisoners had been malnourished for months and even years. And, most just had simple prison shirts and pants, not more than threads at this point, no gear to battle the European winter let alone a cross-country march. Nearly all of these concentration camp inmates had on the German issue wooden clogs as footwear. The German SS had a simple order, anyone who could not keep up was shot. It is not known the exact number of people killed near the war's end on these death marches, but of the estimated 750,000 people ordered to march, it is believed that as many as 350,000 were killed, either by freezing to death, starvation or a German bullet. One Jew from Auschwitz, his name was Ernst Lobethal was among those ordered to march. He only had a few possessions to gather. But a prize possession was a pair of leather boots he had bartered using cigarettes and had hidden for his escape. As he was being ordered to march, he thought this was the time to lace them up.

Denis Avey was a curious and adventurous boy. Born in Essex, just outside London in 1919, he loved to hunt, hike, and explore. In school he was on the boxing team and was a leader among his classmates. He transitioned to Leyton Technical College, and was interested in engineering and learning how things worked. But like most young men born in 1919, a world event would have a profound impact on his life. In 1939, Avey joined the British army.

Avey was sent to North Africa and was part of the British 7th Armored Division known as the Desert Rats. This division was able to push the Italians, who were occupying the area, back out of much of Africa. Then the Germans took command of that theater, a commander named Erwin Rommel. Rommel, who came to be known as the Desert Fox, was smart and aggressive and began to push the British forces back out of North Africa. In 1941, in a counter attack on German forces near Tobruk, Libya, Avey's vehicle was hit. Avey was injured and his best friend, in the passenger seat next to him, was killed. Avey was captured into German hands—he was now a prisoner of war.

Avey's journey to Monowitz prisoner of war camp was a long one. After a long recovery from injuries suffered at Tobruk in a German hospital, Avey escaped. On the run without a plan or ample supplies, he was able to cross the Mediterranean floating on top of a packing crate, but was eventually recaptured. In 1943, after two years of confinement or being on the run, Avey was shipped to a German POW camp called Monowitz. What Avey didn't know, and what few people even know today, is that Monowitz was adjacent to a large industrial complex. Each day POWs would be forced to march to this complex and labored all day working to build this factory, which was a key component in the longer-term German war plans. Working next to these prisoner's of war were Jews from a nearby concentration camp, Auschwitz III.

In his best-selling book titled, *The Man Who Broke Into Auschwitz*, Avey recalls the horrid conditions of these POWs. While the military prisoner's of war endured harsh and tough conditions, Avey recalls that no matter how bad his conditions were, the concentration camps were much worse. Starvation, lack of warm clothing, wooden clogs, brutal treatment—that was their life.

It was dangerous for detainees from the Jewish and POW camps to communicate. German's would brutally beat the Jews if caught communicating with anyone. But Avey befriended someone in the concentration camp. They spoke in code and used alias names so that neither could identify the other by name. In the ongoing communication Avey learned this man had a sister who had escaped to Britain. An area close to where his parents still lived. Using code, Avey wrote to his parents and asked them to find the sister and have her send cigarettes. Cigarettes were the currency in the camps. Cigarettes could buy food and a few other meager necessities. Some time later, Avey received two cartons of cigarettes. He knew who they were from and that he now needed to smuggle them to his friend. Over time, he moved to hand off a pack here and a pack there until the handoff was complete. Avey had always wondered what happened to his friend....

Some five decades later, in 2001, Avey shared his prisoner of war experiences in a radio interview on the subject, including the cigarette handoff. While it may not sound like much today, Avey took much risk to befriend a concentration camp prisoner, send information to his sister, and smuggle cigarettes to him. A reporter heard the radio interview and wondered what Avey wondered, what could have happened to both the sister in Britain and the friend in the camp. After much research the sister was found—her brother? Ernst Lobethal.

While Lobethal had passed away a few years earlier, he lived into his mid-seventies, and attributed the kind acts of a British soldier named 'Ginger' (Avey's code name) with saving his life. You see, he used the cigarettes to trade for boots. Without them, Lobethal does not believe he would have survived the walk.

"You can't live a perfect day," John Wooden said, "Until you do something for someone who will never be able to repay you." One can never compare the risks and kindness people showed each other in our history's darkest moments. But, there is a theme that carries forward into our daily safety. Safety, maybe like no other aspect of today's work life, offers us the ability to have a profound, lasting and sometimes life saving effect on another person. A very small act of kindness, like reminding someone of a safety rule or giving one a pair of safety glasses, or other PPE, can literally save a life. Maybe if we care as Avey did, some five decades from now you will learn that something we did saved a life. Why not plant those seeds today?

Source:
Avey, Denis, Rob Broomby, *The Man Who Broke Into Auschwitz: A True Story of World War II*, Da Capo Press, 2011.

48 Oh Yeah, and Then There Is Safety Leadership!

It was a typical October day for Granite City high school, located near St. Louis, Missouri. When school ended, several members of the lady's soccer team piled in the back seat of the coach's car and headed toward their afternoon game. The only problem was that on the short trip to the game the coach checked into his Twitter account on his smart phone—while driving—70 mph down the interstate!

This concerned the student athletes in the back seat so much that one of them snapped a picture of the coach driving while reading his phone. The student later shared her concern and picture with a parent. The photo was shared with other parents who quickly became concerned with their children's safety.

Parents moved forward to contact Jim Greenwald, the school's superintendent. Illinois state police say this act of driving while reading one's phone is illegal.

As a parent, this is a very troubling example of putting people at risk with a mindless act. But, shortcuts happen every minute in the workplace. A May 2010 *Business Week* article entitled "The Peer Principle" said, "In the area of safety, our study found that 93 percent of employees say they see urgent risks to life and limb..."

How we lead in safety can make a big difference in eliminating shortcuts such as the one the soccer coach took and the ones our employees are taking. Here are four tips to LEAD safety.

L: Look at Your Shadow. Each one of us casts a leadership shadow. It is a reflection of what we do and what we say. The shadow is cast over those who are watching us at any given moment in time. In the case of the soccer coach, it was his student athletes in the back seat. In your family, it may be your children, spouse, and/or parents. And, in the workplace it is your coworkers. From time to time, we need to step back and ask "what shadow am I casting?" I'm sure the soccer coach would cast a different shadow if given the opportunity. What does your shadow say about your leadership?

E: Engage in Crisis. There is an old saying, "If there are no good rumors, make one!" This is not what I am suggesting. Instead, engaging in crisis points to the fact that leadership in times of calm and routine is easier, but decisions in times of crisis are what your people will remember. It doesn't matter if it is a small crisis, such as a conversation with a crew on a safety rule, and following that rule will mean the job lasts longer. Or, in the electric industry, it could be a restoration effort after a major storm that has left tens of thousands of customers out of power. In this case, the entire organization is watching for safe restoration vs. timely restoration. "Decisions are easy when values are clear," Roy Disney, nephew of Walt Disney, said. Values are best shown in crisis—what are your safety values?

A: Align Your Words and Actions. Being a father to a teenage girl and an 11-year-old boy, one of my favorite quotes is from Robert Fulghum and it says, "Don't worry that children never listen to you; worry that they are always watching you." Students were watching their soccer coach. "I think this was a onetime only case in which he was wrong, we brought it to his attention that he was wrong, he knows that he was wrong, and it will not happen again," Superintendent Jim Greenwald said. I'm sure the coach would have told students to never be on the phone when driving. "Don't do what I do but do what I say" doesn't work. Who is watching you? And, do your words align with your actions?

D: Decide to Lead. John Maxwell, bestselling author on leadership, has simply defined a leader as, "someone with influence over another." The fact is that we are all leaders at any given point in time. At any time someone is watching you, what you are doing, how you are doing it, what you are wearing and how you are carrying yourself. In fact, the soccer coach was a leader when he drove students athletes to a game while checking out Twitter at 70 mph. Leadership understands that people are watching; good leadership simply means that we have decided to influence in a positive and powerful direction.

Warren Bennis, an American scholar, organizational consultant and author, and widely regarded as a pioneer of the contemporary field of leadership studies, said, "The most dangerous leadership myth is that leaders are born—that there is a genetic factor to leadership. This myth asserts that people simply either have certain charismatic qualities or not. That's nonsense; in fact, the opposite is true. Leaders are made rather than born." Have you decided to lead?

Source:
Grenny, Joseph, "The Peer Principle," *Bloomberg Businessweek—The Influential Leader,* May 2010.

In Closing

> *"Every improvement is a sign of change, but not every change is a sign of improvement."* —Anonymous

First, thank you for reading *What Utility Safety Leaders Do*!

I'm actually buttoning up all the loose ends of this book such as final editing, cover design, and layout at the end of the year. It's the Christmas holidays. Our house is filled with decorations, family, food, and cheer.

In particular the holidays bring extra focus to family and friends—the people that we love and depend on, and the people who depend on us. The holidays are a great time to reconnect, remember, reminisce, and be truly thankful for our blessings. Due to emphasis on family it is also a stark reminder of what some of our utility brothers and sisters have lost. Losses due to life changing workplace incidents. And unfortunately, workplace fatalities. Utility workers who went to work one day and never returned home.

There is an old saying, "Our best thinking got us here." To me this saying means that we have done all that we can do and pushed things as hard as we can push them . . . but our work is not finished. To be even more effective we must improve the current status quo. And, to make improvements and get even better results we must change.

The holidays remind us of what's important, but safety is important each and every day. Each day families let their moms and dads go to work and they expect us to keep them safe so that they return home. That's our job. This book can help.

This book is offered from one utility guy to another in the sincere spirit that you captures a number of new ideas, ideas, tips, tactics, and concepts that you can take off these pages and place in your safety meetings or crews. Whether you are influencing one person in your organization or thousands, I sincerely hope this book will serve as a valuable resource.

Thanks for reading. And never forget, readers are leaders . . . so now it's time to take these new ideas and lead.

Wishing you the best. And remember, *No One Gets Hurt Today*. —Matt

About the author

Matt Forck is a certified safety professional (CSP) and a former journey electrical lineman with over two decades of experience in the hazardous field of electricity distribution. Matt is founder and director of SafeStrat, a boutique safety keynote and safety consulting services organization, providing dynamic and tailored presentations, training, and consulting services to clients throughout the United States. Matt resides in Columbia, Missouri with is wife and two children. Learn more about Matt and Safestrat at www.safestrat.com.

www.ingramcontent.com/pod-product-compliance
Lightning Source LLC
Chambersburg PA
CBHW051651170526
45167CB00001B/421